Communications
in Computer and Information Science   1854

Rationale

The CCIS series is devoted to the publication of proceedings of computer science conferences. Its aim is to efficiently disseminate original research results in informatics in printed and electronic form. While the focus is on publication of peer-reviewed full papers presenting mature work, inclusion of reviewed short papers reporting on work in progress is welcome, too. Besides globally relevant meetings with internationally representative program committees guaranteeing a strict peer-reviewing and paper selection process, conferences run by societies or of high regional or national relevance are also considered for publication.

Topics

The topical scope of CCIS spans the entire spectrum of informatics ranging from foundational topics in the theory of computing to information and communications science and technology and a broad variety of interdisciplinary application fields.

Information for Volume Editors and Authors

Publication in CCIS is free of charge. No royalties are paid, however, we offer registered conference participants temporary free access to the online version of the conference proceedings on SpringerLink (http://link.springer.com) by means of an http referrer from the conference website and/or a number of complimentary printed copies, as specified in the official acceptance email of the event.

CCIS proceedings can be published in time for distribution at conferences or as post-proceedings, and delivered in the form of printed books and/or electronically as USBs and/or e-content licenses for accessing proceedings at SpringerLink. Furthermore, CCIS proceedings are included in the CCIS electronic book series hosted in the SpringerLink digital library at http://link.springer.com/bookseries/7899. Conferences publishing in CCIS are allowed to use Online Conference Service (OCS) for managing the whole proceedings lifecycle (from submission and reviewing to preparing for publication) free of charge.

Publication process

The language of publication is exclusively English. Authors publishing in CCIS have to sign the Springer CCIS copyright transfer form, however, they are free to use their material published in CCIS for substantially changed, more elaborate subsequent publications elsewhere. For the preparation of the camera-ready papers/files, authors have to strictly adhere to the Springer CCIS Authors' Instructions and are strongly encouraged to use the CCIS LaTeX style files or templates.

Abstracting/Indexing

CCIS is abstracted/indexed in DBLP, Google Scholar, EI-Compendex, Mathematical Reviews, SCImago, Scopus. CCIS volumes are also submitted for the inclusion in ISI Proceedings.

How to start

To start the evaluation of your proposal for inclusion in the CCIS series, please send an e-mail to ccis@springer.com.

Ana Fred · Carlo Sansone · Kurosh Madani
Editors

Deep Learning Theory and Applications

First International Conference, DeLTA 2020
Virtual Event, July 8–10, 2020, and Second International
Conference, DeLTA 2021, Virtual Event, July 7–9, 2021
Revised Selected Papers

 Springer

Editors
Ana Fred
University of Lisbon
Lisbon, Portugal

Instituto de Telecomunicações
Lisbon, Portugal

Kurosh Madani
Université Paris-Est Créteil (UPEC)
Créteil, France

Carlo Sansone
Università degli Studi di Napoli
Napoli, Italy

ISSN 1865-0929 ISSN 1865-0937 (electronic)
Communications in Computer and Information Science
ISBN 978-3-031-37319-0 ISBN 978-3-031-37320-6 (eBook)
https://doi.org/10.1007/978-3-031-37320-6

This Springer imprint is published by the registered company Springer Nature Switzerland AG
The registered company address is: Gewerbestrasse 11, 6330 Cham, Switzerland

Preface

The present book includes extended and revised versions of a set of selected papers from the 1st and 2nd International Conferences on Deep Learning Theory and Applications (DeLTA 2020 and DeLTA 2021), which were exceptionally held as online events due to COVID-19. DeLTA 2020 was held from 8 to 10 July 2020, and DeLTA 2021 was held from 7 to 9 July 2021.

DeLTA 2020 received 28 paper submissions from 17 countries, of which 14% were included in this book. DeLTA 2021 received 30 paper submissions from 18 countries, of which 13% were included in this book.

The papers were selected by the event chairs and their selection is based on a number of criteria that include the classifications and comments provided by the program committee members, the session chairs' assessment and also the program chairs' global view of all papers included in the technical program. The authors of selected papers were then invited to submit a revised and extended version of their papers having at least 30% innovative material.

Deep Learning and Big Data Analytics are two major topics of data science, nowadays. Big Data has become important in practice, as many organizations have been collecting massive amounts of data that can contain useful information for business analysis and decisions, impacting existing and future technology. A key benefit of Deep Learning is the ability to process these data and extract high-level complex abstractions as data representations, making it a valuable tool for Big Data Analytics, where raw data is largely unlabeled. Machine learning and artificial intelligence are pervasive in most real-world application scenarios such as computer vision, information retrieval and summarization from structured and unstructured multimodal data sources, natural language understanding and translation, and many other application domains. Deep learning approaches, leveraging on big data, are outperforming state-of-the-art more "classical" supervised and unsupervised approaches, directly learning relevant features and data representations without requiring explicit domain knowledge or human feature engineering. These approaches are currently highly important in IoT applications.

The papers selected to be included in this book contribute to the understanding of relevant trends of current research on Deep Learning Theory and Applications, including: Generative Adversarial Networks (GAN), Convolutional Neural Networks (CNN), Object Detection, Sparse Coding, Semantic Segmentation, Learning Deep Generative Models, Human Pose Estimation, Clustering, Classification and Regression, Autoencoders, Recurrent Neural Networks (RNN), Image Classification, Graph Representation Learning, Gaussian Processes for Machine Learning and Dimensionality Reduction.

We would like to thank all the authors for their contributions and also the reviewers who have helped to ensure the quality of this publication.

July 2021

Ana Fred
Carlo Sansone
Kurosh Madani

Organization

Conference Chair

Kurosh Madani   University of Paris-Est Créteil, France

Program Co-chairs

2021

Carlo Sansone   University of Naples Federico II, Italy

2020 and 2021

Ana Fred   Instituto de Telecomunicações and University of
                                Lisbon, Portugal

Program Committee

Served in 2020

Alessandro Artusi	RISE Ltd, Cyprus
Andrew Reader	Kings College London, UK
Bernadette Dorizzi	Télécom SudParis, France
Carlo Sansone	University of Naples Federico II, Italy
Chih-Chin Lai	National University of Kaohsiung, Taiwan, Republic of China
George Awad	National Institute of Standards and Technology, USA
Hasan Bulut	Ege University, Turkey
Li Shen	Icahn School of Medicine at Mount Sinai, USA
Liangjiang Wang	Clemson University, USA
Marcilio de Souto	University of Orléans, France
Marco Turchi	Fondazione Bruno Kessler, Italy
Matias Carrasco Kind	University of Illinois Urbana-Champaign, USA
Mihail Popescu	University of Missouri, USA

George Papakostas	International Hellenic University, Greece
Gilles Guillot	CSL Behring/Swiss Institute for Translational and Entrepreneurial Medicine, Switzerland
Huaqing Li	Southwest University, China
Juan Pantrigo	Universidad Rey Juan Carlos, Spain
Ke-Lin Du	Concordia University, Canada
Luis Anido-Rifon	University of Vigo, Spain
Nebojsa Bacanin	Singidunum University, Serbia
Perry Moerland	Amsterdam UMC, University Amsterdam, The Netherlands
Ryszard Tadeusiewicz	AGH University of Science and Technology, Poland
Seokwon Yeom	Daegu University, South Korea
Stefano Cavuoti	University of Naples Federico II, Italy
Stephan Chalup	University of Newcastle, Australia
Sunghwan Sohn	Mayo Clinic, USA
Yang Liu	Harbin Institute of Technology, China
Yizhou Yu	University of Hong Kong, China
Yung-Hui Li	National Central University, Taiwan, Republic of China

Additional Reviewers

Served in 2020

Chaowei Fang	University of Hong Kong, China
Haoyu Ma	University of Hong Kong, China

Served in 2021

Bingchen Gong	University of Hong Kong, China

Invited Speakers

2020

Petia Radeva	Universitat de Barcelona, Spain
Vincent Lepetit	École des Ponts ParisTech, France

2021

Matias Carrasco Kind	University of Illinois Urbana-Champaign, USA
Fabio Scotti	Universita degli Studi di Milano, Italy
Elisabeth André	University of Augsburg, Germany
Petia Radeva	Universitat de Barcelona, Spain

Contents

Alternative Data Augmentation for Industrial Monitoring Using
Adversarial Learning 1
   Silvan Mertes, Andreas Margraf, Steffen Geinitz, and Elisabeth André

Multi-stage Conditional GAN Architectures for Person-Image Generation 24
   Sheela Raju Kurupathi, Veeru Dumpala, and Didier Stricker

Evaluating Deep Learning Models for the Automatic Inspection
of Collective Protective Equipment 49
   Bruno Georgevich Ferreira, Bruno Gabriel Cavalcante Lima,
   and Tiago Figueiredo Vieira

Intercategorical Label Interpolation for Emotional Face Generation
with Conditional Generative Adversarial Networks 67
   Silvan Mertes, Dominik Schiller, Florian Lingenfelser, Thomas Kiderle,
   Valentin Kroner, Lama Diab, and Elisabeth André

Forecasting the UN Sustainable Development Goals 88
   Yassir Alharbi, Daniel Arribas-Bel, and Frans Coenen

Disrupting Active Directory Attacks with Deep Learning for Organic
Honeyuser Placement 111
   Ondrej Lukas and Sebastian Garcia

Crack Detection on Brick Walls by Convolutional Neural Networks Using
the Methods of Sub-dataset Generation and Matching 134
   Mehedi Hasan Talukder, Shuhei Ota, Masato Takanokura,
   and Nobuaki Ishii

Author Index 151

Alternative Data Augmentation for Industrial Monitoring Using Adversarial Learning

Silvan Mertes[1], Andreas Margraf[2(✉)], Steffen Geinitz[2], and Elisabeth André[1]

[1] University of Augsburg, Universitätsstraße 1, 86159 Augsburg, Germany
{silvan.mertes,elisabeth.andre}@informatik.uni-augsburg.de
[2] Fraunhofer IGCV, Am Technologiezentrum 2, 86159 Augsburg, Germany
{andreas.margraf,steffen.geinitz}@igcv.fraunhofer.de

Abstract. Visual inspection software has become a key factor in the manufacturing industry for quality control and process monitoring. Semantic segmentation models have gained importance since they allow for more precise examination. These models, however, require large image datasets in order to achieve a fair accuracy level. In some cases, training data is sparse or lacks of sufficient annotation, a fact that especially applies to highly specialized production environments. Data augmentation represents a common strategy to extend the dataset. Still, it only varies the image within a narrow range. In this article, a novel strategy is proposed to augment small image datasets. The approach is applied to surface monitoring of carbon fibers, a specific industry use case. We apply two different methods to create binary labels: a problem-tailored trigonometric function and a WGAN model. Afterwards, the labels are translated into color images using pix2pix and used to train a U-Net. The results suggest that the trigonometric function is superior to the WGAN model. However, a precise examination of the resulting images indicate that WGAN and image-to-image translation achieve good segmentation results and only deviate to a small degree from traditional data augmentation. In summary, this study examines an industry application of data synthesization using generative adversarial networks and explores its potential for monitoring systems of production environments.

Keywords: Image-to-Image translation · Carbon fiber · Data augmentation · Computer vision · Industrial monitoring · Adversarial learning

1 Introduction

1.1 Motivation

Visual inspection respresents a wide spread methodology in industrial quality control which is usually employed in mass production to ensure quality standards. With recent progress in deep learning research, the focus of computer vision shifted from image processing filters to neural network architecture selection and design. The increasing level of automation and digitalisation in the manufacturing industry has drawn attention to sensing technology and camera sensors in particular. The field of online process monitoring (OPM) primarily deals with imaging technology to detect changes, faults or potential anomalies in continuous production environments. Of course, the actual bene-fit of measurement technology depends on its level of automation. Therefore, intelligent image processing is a key feature of monitoring systems.

In recent years, machine learning algorithms have progressively overtaken filter based image processing, a fact that has been discussed in relevant publications [4,25]. In the context of computer vision, convolutional neural networks (CNNs) have proven to be superior because they generalize better. Given large training and sample data, they are flexible across domains and therefore can be applied to very different kinds of applications [15,35].

1.2   Pushing the Limits of Image Segmentation

Highly specialized industries are often confronted with incomplete or insufficient data. With increasing effort spent on data collection and preparation - tasks that require time and skilled personnel - deep learning models become inefficient, expensive and there-fore unattractive. ML solutions only serve the cause if they add to productivity. This article discusses concepts that augment scarce datasets and allow semantic segmenta-tion models to achieve a high accuracy when trained on this data.

Research work has been conducted in the field of data preparation, cleaning and augmentation using e. g. algorithms for interpolation, smoothing, simple transformation or filtering [33]. Although these methods help to create the precondition for successful model development, they quickly appear constrained in their variation space. This arti-cle takes a closer look to a branch of research exploring the representation of artificially generated data based on real world blueprints, an approach also denoted *synthetic data generation*. In this article we consider semantic segmentation on carbon fiber textiles with unique surface structures and heterogeneous anomalies. Earlier publications have shown that image processing based on conventional filters such as edge, contour, thresh-old or Fourier transformation do not cover anomalies with the desired accuracy. In this respect, we offer a CNN-based method using two different models for adversarial learn-ing to augment the data. Thus, our DA methods are based on trigonometric functions, *Wasserstein Generative Adversarial Networks* (WGANs) and *pix2pix* image-to-image translation. As an overall goal, we pursue this approach to improve the reliability of defect detection and reduce the costs of model training by automating a considerable portion of the data preparation tasks.

One of the main advantages of CNNs is their ability to classify images, i.e. to pre-dict which category an image, object within the image or single pixel belong to. This dependency is usually determined on the basis of the "euclidean distance", a widely used similarity score. The latest generation of CNNs requires large amounts of samples and data to achieve reasonable results. Manual annotation cannot be fully automated to this date, but still constitutes a time consuming task and - since it is performed by peo-ple - is prone to error over longer periods of time. Regular DA has been a popular and obviously simple tool to increase the number of training samples. DA methods include rotation, scaling, blurring and similiar transformations. These functions adjust image and label pairs carefully and to a limited extent but do not exploit the full potential that is offered by *Generative Adversarial Networks* (GANs) and image-to-image translation.

GANs are designed to output 'new' images which appear to be as realistic as possible, so that they are indistinguishable from real-world photographs. Image-to-image translation as performed by *pix2pix* improves the ability of the original GANs to transfer images from one domain to another. However, input data for GAN training should be carefully collected or created with the expected application domain in mind. In general, there are two possible approaches to provide the input. One very obvious approach would be to gather data which already contain the desired image-to-image translation information. This could be accomplished by e. g. acquiring images from different angles using the same camera or by moving the target around. This has been examined for the fashion industry in recent publications [7, 19].

In the manufacturing industry such intensive interfering with the process is undesirable because it may cause time delays or unscheduled interruptions. Therefore, getting access to the target from more than one angle may not be possible. As an alternative approach, one can use simluated data or approximate image-to-image translation by using a specifically designed algorithm. Of course, the second suggestion can only be applied to smaller sets of data in order to keep the workflow efficient.

In this article, image-to-image translation is used to translate randomly generated binary labels to images. The approach allows for artificially creating label and image pairs that are actually new and serve as training samples for semantic segmentation models. For the generation of new labels, we propose two distinct concepts: the first approach is based on a function precisely tailored to the application and was already introduced by Mertes et al. [27]. The second one is presented the first time in this paper and uses a WGAN model trained on the original binary labels in order to allow the generation of synthetic labels. In both cases, binary label images are generated and used for further processing, i.e. for generating image/label pairs by using an image-to-image translation system. Designing problem-specific functions requires good domain understanding and further effort to model an algorithm and tune it to adjust to the given problem. The resulting function is transparent and human-readable which allows for better debugging and testing. However, the proposed WGAN-based approach requires less manual effort and automates a considerable part of the overall process.

All in all, we present a novel approach, to augment image data for semantic segmentation networks by applying image-to-image translation with both, a domain-specific mathematical model and an approach entirely based on generative models. We test both approaches based on images containing carbon fiber surface defects and discuss the results.

1.3 Structure

The structure of this article is divided into four main sections: at first, we provide an overview on related work and existing technology in Sect. 2. The subsequent section present the two approaches (Sect. 3) followed by the experimental setup (Sect. 4). We then discuss relevant results and critically reflect them in respect of related concepts in Sect. 5. The final section covers concluding statements based on our findings and provides an outlook on future work as can be seen in Sect. 6.

2   Related Work

This section lists and discusses previously published papers in related research areas which include Machine Learning (ML), Artificial Neural Networks (ANN), Computer Vision (CV), GANs, OPM and Organic Computing (OC) as well as publications that contributed to the research presented in this article. The identification of anomalies on carbon fibers, e.g. the misalignment on textile surfaces, has been discussed in various publications [11, 12]. In the same context, Margraf et al. examined the self-adaptation of image processing filters using organic computing paradigms [22]. Methodologies for automated algorithm selection and filter pipelines have been discussed by Stein et al. [38]. A partly self-adaptive algorithm for data analysis based on carbon fiber monitoring was introduced by Margraf et al. [21].

The domain of artificial neural networks took large steps forward when AlexNet [18], GoogleNet [40] and VGGNet [35] were introduced for the classification of large image sets. Several publications address industrial monitoring applications: Masci et al. used CNN for classification of steel defects, and Soukup et al. used CNN for photometric stereoscopic images [24, 36]. A region proposal network for real-time object detection was presented by Ren et al., while Ferguson et al. used CNNs and transfer learning to detect X-ray image defects [9, 29]. Furthermore, the use of CNNs for industrial surface anomaly inspection was explored by Staar et al. [37]. The first ones to introduce pixel-based segmentation was Long et al. [20], while Schlegl et al. published a work in which GANs for marker detection were used for unsupervised anomaly detection [32]. A survey exploring GANs for anomaly detection was presented by Di Mattia et al. [8]. Methods for color translation between different photographic contexts were discussed in various related publications [41, 43]. Translation of completely different image domains was first achieved by Isola et al., who presented the *pix2pix* architecture [17]. The *pix2pix* architecture was the first to allow the projection of various image domains such as edge objects or label images to colored photographs. A huge problem when dealing with pixel-based segmentation tasks is a big gap between a small foreground and a relatively large background. This is especially reflected in the case of carbon fiber surface images. Ronneberger et al. confirmed, that multi-channel feature maps prove to be more useful in these tasks [31]. It should be noted that this architecture makes heavy use of data enhancement, while taking only little input data. However, its capabilities are limited when dealing with very small training sets. Multiple approaches to use GANs for the specific purpose of data augmentation were presented in different works. Frid-Adar et al. and Mariani et al. were able to apply GANs to classification tasks, while Choi et al. presented an approach that makes use of image-to-image translation networks for semantic segmentation tasks by transforming labeled data to related image domains so that the original label still fits to the artificially created image [5, 10, 23]. Huang et al. utilized image-to-image networks for shared segmentation tasks by using multiple image domains for the training process [16]. The authors of this paper could not identify any publication dedicated to the use of GANs for generating entirely new image and label pairs for enhanced training datasets in the context of semantic image segmentation.

An early application of generative models was presented by Pathak et al. for image inpainting to reconstruct a region within an image that adapts well to its surroundings

[28]. Shrivastava et al. proposed an approach denoted Simulated+Unsupervised (S+U) learning which allows to improve the output of a simulator based on a GAN network [34]. Furthermore, *StarGAN* was introduced by Choi et al. to improve the quality of translated images by providing a more scalable method for image-to-image translation to an arbitrary domain [6].

The WGAN architecture constitutes a variant of generative models. It was introduced to increase learning stability and is less prone to mode collapse. Its structure ressembles the general GAN approach, except for its discriminator that learns on the basis of the Wasserstein distance by approximating a 1-Lipschitz function [2]. Later, Arjosvky et al. made further adjustments to the weight clipping which improved the WGAN behaviour to better handle hyperparameter tuning [13]. In addition, a comparative study increased the understanding of GAN behaviour regarding training stability and model saturation [1]. The authors of this article are aware that metaheuristics for hyperparameter optimization, e. g. swarm intelligence [39] exist. However, this field of research is not subject to this research article.

3   Approach

The following sections explain the concepts that are introduced in this paper. The main idea of our approach is that we are enhancing datasets with augmented data by artificially modeling label images and after that convert them into real image data. By doing so, we get image/annotation pairs that are needed for the training of neural networks for semantic segmentation tasks. To create new label data, we propose two different methods. The first method, that we already introduced in [27], is an algorithm specifically designed for our particular application at hand, i.e. the defect detection in carbon fiber structures. It is based on a randomized label generator that uses a stochastically parametrized function to build segmentation masks. The second method is a more generic one. It uses a WGAN that is trained on raw segmentation masks. After training, the WGAN is capable of generating new label images that appear similar to the original labels. By applying this concept, we get rid of the engineering overhead that is necessary when using the first method. While the randomized label generator that is used by the first method has to be defined and optimized specifically for every new segmentation task, the WGAN should be able to learn the label structure of new tasks by itself.

The labels that are produced by either of the two methods are fed into a *pix2pix* network that was trained on an image-to-image conversion task, i.e. the network was trained to convert label mask images to real images that fit to the respective labels, thus resulting in image/label pairs that can be used to enlarge training datasets.

All in all, our approach can be seen as a three-folded system: first, we train a *pix2pix* network on an image-to-image translation task, so that it learns to perform a translation from labels of defect images to their corresponding image data. Second, one of the aforementioned methods is used to generate synthetic label data. At last, the synthetic label data is fed into the trained *pix2pix* network, resulting in new training pairs for further machine learning tasks. The following sections explain these steps in more detail.

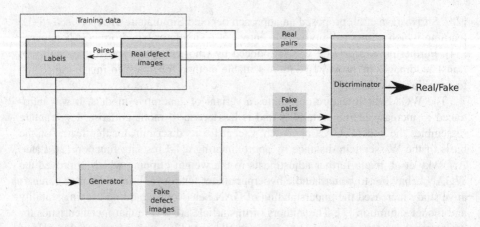

Fig. 1. Training of a *pix2pix* network to perform image-to-image translation between labels and defect images (Step 1) [27].

3.1 Label-to-Image Model

In order to convert label masks to corresponding images, we trained a *pix2pix* model. For the training, a dataset of existing real defect images and manually labeled annotation masks was used. The basic scheme of the training process is depicted in Fig. 1. We used the *pix2pix* network architecture that was introduced by [17]. We adapted the size of the input layer to fit the dimensions of our dataset. Other than that, we did not make any modifications to the originally proposed architecture.

3.2 Synthetic Label Generation

The idea of our approach is to feed new synthetic label data into the *pix2pix* network in order to obtain new pairs of defect images and label masks. As mentioned above, two different methods were applied for the stage of synthesizing new label data.

Mathematical Modelling of Defects. The first method is based on the observation that in many application scenarios label masks have a common structure as proposed by Mertes et al. in their related publication [27]. It is illustrated in Fig. 2. The idea behind this first approach to generate fake label masks is to find a mathematical description of those structures for a specific case. In the application scenario that serves as an example for evaluation in this article - the mentioned defect detection on carbon fiber structures - label masks usually appear as mostly straight or curved lines. Those structures can be seen as a combination of multiple graphs with different rotations and varying thickness of the plots. Thus, the mathematical description of a single defect label could be approximated through a trigonometric function. By adding a stochastic factor to such a function, we can plot different graphs that can be considered as new, artificial label masks. For our specific task, we conducted several experiments that showed that the following function $f(x)$ can be used to cover a huge part of carbon fiber defect structures. We denote $f(x)$ as already presented in [27]:

$$f(x) = a_1 \cdot sin(a_2 \cdot x) + a_3 \cdot sin(x) + a_4 \cdot cos(a_5 \cdot x) + a_6 \cdot x + a_7 \cdot x^2$$

where the parameters a_n are chosen randomly within certain defined intervals. For our specific experiments, we found appropriate intervals by visual exception of carbon fiber defect images. By using those intervals, that are listed in Table 1, we ensure to cover a wide range of different defect structures. To that end, a sine function was tuned with a rather big amplitude to model the global structure of the label, that is typically shaped in curves. For the structure on a more micro perspective level, we used another sine with a much smaller amplitude interval. Aperiodic curvings were modeled by the use of polynomic functions. As described in [27], we randomly set the variables and plotted the resulting graph for every fake label for $x \in [0, w]$ where w represents the width of the sample images. After creating those plots they were rotated randomly. At last, we took a random number of those graphs, randomized the thickness of the resulting lines, and overlapped those graphs to create images of labels with a realistic fiber-like appearance.

The authors are aware of the fact that this method is very specific to the given task at hand. A lot of engineering time and effort has to be spent to find sufficient mathematical models for the respective task. However, similar approaches for defect-modeling have been successfully applied to similiar problems in previous work [14].

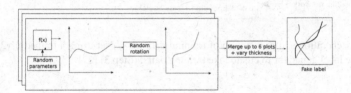

Fig. 2. Heuristic to generate fake labels using the label generator (Step 2) [27].

Table 1. Parameters for the fake label generator [27].

Parameter	Lower bound	Upper bound
a_1	15	30
a_2	0.02	0.03
a_3	1	50
a_4	−0.5	0.5
a_5	−0.5	0.5
a_6	−0.5	0.5
a_7	0.005	0.0095

Generative Modelling of Defects. The fact that the approach of mathematically modeling label mask structures is coupled to a lot of engineering overhead led to the investigation of a more generic approach, which is described in this section.

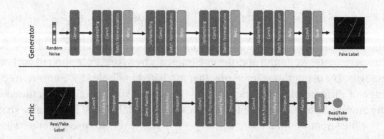

Fig. 3. Architecture of both the generator and critic of the used WGAN network.

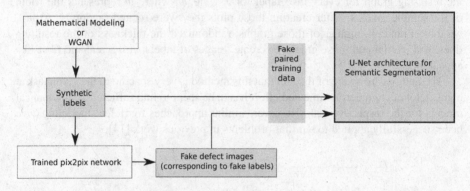

Fig. 4. The generation and preparation of training data for U-Net using a trained *pix2pix* model and the fake label generator to create fake training pairs. (Step 3) [27].

The basic idea of this method is to use the capability of original GANs to transform random noise vectors into data that looks similar to data of a given training set. While the pix2pix network that was described earlier in this work performs a transformation between different image domains, more traditional GANs are designed to generate completely new data. This property was used for data augmentation tasks in the past, not only in the image domain [3], but also for audio classification tasks [26]. However, instead of generating new image data of carbon fiber defect images, our approach uses such a rather traditional GAN to create new label segmentation masks, which then can be transformed to defect image data by feeding it to our pix2pix network, as will be described in the next section. The benefit of generating label masks instead of real image data is that, on the one hand, the pixelwise structure of those label masks appears less complex due to its binary nature. On the other hand, the cascade of generating label masks in one step and transforming those label masks further to actual image data in another step makes those image/label pairs available for direct processing: they finally serve as a training set for segmentation networks.

To generate those artificial label mask images, we make use of a convolutional GAN that operates on the *Wasserstein-Loss* as introduced by Arjovsky et al. [2]. The network architecture of both the generator and the *critic*, as the discriminator is called in the context of WGANs, is illustrated in Fig. 3.

As a training dataset, we use real label masks. More specifically, the same label masks can be used which were already part of the training pairs of the *pix2pix* network. Details regarding the training configuration can be found in Sect. 4.5.

3.3 Finalizing the Training Data

In the last step, the generated label data is used to create new corresponding image data. Thus, the label data that was produced by either mathematical modelling or by the WGAN is used as input to the trained *pix2pix* model. The resulting data pairs of label/image data can be used to train further networks for the actual segmentation task. The whole system is shown in Fig. 4. For our experiments, we chose a U-Net architecture to perform this segmentation task. It has to be emphasized that the selection of this specific network was done for the purpose of evaluating and comparing our augmentation approaches, and that the authors don't claim that architecture to be the best choice for the respective task. However, U-Net could achieve promising results in related fields like biomedical image segmentation [31].

Fig. 5. Examples of real image data pairs labelled by experts. The misaligned fibers are visible on top of the fiber carpet [27].

4 Experiments and Discussion

4.1 Dataset Specifics

Our system was evaluated in the context of an industrial application scenario. More precisely, the domain of carbon fiber defect monitoring was chosen for testing and evaluating of the proposed approaches. In images of fiber structures without recognizable defects, the single fibers are aligned in parallel and form a carpet of straight lines. During the production process, mechanical stress caused by spools in the transportation system can lead to damage of the fiber material, which usually can be recognized as misaligned fibers. The shape of those cracked fibers, as well as their position and size vary heavily. Thus, there is no *template* for single defects. Given the different images of defective fiber material, a huge variety of defect structures can be observed. In this specific use case, we aim for the identification of defects on a carbon fiber carpet. To achieve this goal, a U-Net architecture is trained to perform a binary segmentation of pixels that contain defects. Figure 5 shows examples of defect images with corresponding binary labels. The environment and the design of the monitoring system that was used to acquire the image data for our experiments has been described in earlier publications [11,12,22].

4.2 Experimental Setup

For a meaningful evaluation, we ran several experiments to compare the two variants of our approach with conventional data augmentation methods. Thus, parts of our datasets that are described below were augmented with traditional data augmentation techniques. The following *simple* image transformations were applied to those artificially extended datasets:

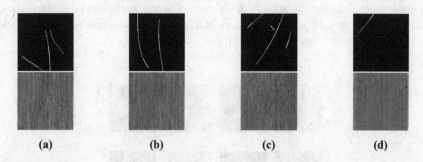

    (a)   (b)   (c)   (d)

Fig. 6. Samples of synthetic labels (top row) and corresponding *pix2pix* outputs (bottom row) imitating misaligned fibers [27].

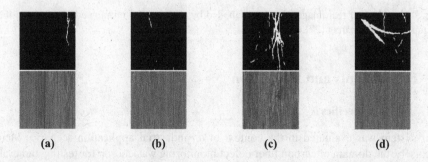

    (a)   (b)   (c)   (d)

Fig. 7. Samples of synthetic labels generated by WGAN (top row) and corresponding outputs (bottom row) using the same *pix2pix* model as for Fig. 6.

- Randomised crop of squares of different size (*RandomSizedCrop*)
- Horizontal and vertical flip
- Rotation (for 180°)
- Elastic transformation
- Grid distortion

We arranged the image data in six different sets and performed multiple trainings of a U-Net architecture. Then, we used the resulting models to make predictions on a test set. To ensure comparability, the same test set was used for every training set. Every training pair for the U-Net architecture consists of a real or fake defect image and a real or fake binary label image. The first four datasets are the exact same as used in the publication by Mertes et al. [27]. We added two additional sets using the WGAN model for generating fake labels as listed below:

Dataset 1. contains 300 pairs of real defect data and corresponding binary label images. Thus, only original data without DA was taken.

Dataset 2. contains the same 300 pairs of defect data and corresponding labels as *dataset 1*. Additionally, an *online* form of data augmentation was applied as described above. For each image some of those aforementioned transformation operations were performed with a predefined probability.

Dataset 3. contains 3000 pairs of defect data and corresponding labels. 2700 of the 3000 data pairs were generated by applying the *pix2pix* based data augmentation approach on *dataset 1*. For the creation of synthetic label, our mathematical model was applied. Furthermore, the 300 original data pairs from *dataset 1* were taken.

Dataset 4. contains the same 3000 pairs of defect data and corresponding labels as *dataset 3*. Additionally, the same conventional stochastic data augmentation as for *dataset 2* was applied, i.e. each image was transformed with a predefined probability during training. Thus, *dataset 4* combines common data augmentation with our approach.

Dataset 5. contains 3000 pairs of defect data and corresponding labels. In this dataset, 2700 of the 3000 pairs were generated by the *pix2pix* network. This time, however, the input data for image-to-image-translation were not generated from a trigonometric function, but by training a *WGAN* model on image pairs of real sample data. The resulting model was then used to create new binary labels. The remaining 300 image pairs were taken from *dataset 1* as performed in the previous datasets.

Dataset 6. contains the same 3000 pairs of defect data and corresponding labels as *dataset 5*. Hereby, though, we altered the training of the U-Net by adding *online, stochastic* data augmentation as for *dataset 4* and *dataset 2*. In this dataset, 2700 of the 3000 pairs were generated by the *pix2xpix* network. This was performed to examine how *regular* data augmentation will change the result on top of WGAN based label generation and image-to-image translation.

Each of the datasets was used to train a separate U-Net model for semantic segmentation. For testing and evaluation, a single, distinct dataset was used, containing real defect data and annotations that are provided by domain experts.

4.3 Pix2Pix Configuration

The configuration of the *pix2pix* model is given in Table 2. We stopped the training after 3200 epochs, as we could not observe any further improvement of the generated images by that time. Figure 6 shows a selection of pairs of labels and images generated through application of the *pix2pix* model, where the labels where created by mathematical modeling. Figure 7 shows label/image pairs where the labels were generated by our WGAN approach.

Table 2. Pix2Pix Configuration [27].

Parameter	Value
Learning rate	0.0005
Batch Size	1
Epochs	3200
Loss Function	Mean Squared/Absolute Error

4.4 U-Net Configuration

The U-Net architecture was trained individually for every dataset. As described above, dataset 2, 4 and 6 were augmented with traditional DA, i.e. conventional image transforms.

All of those conventional data augmentation methods are based on the library published by [30]. We used the same methods that were already described in [27]. A stochastic component was added to the image transformations, i.e. all operations were performed with a given probability.

The randomized crop was given the probability $p = 0.25$ and a window size interval of $[400, 512]$ pixels. Furthermore, the probabilities for flipping, rotation, elastic transform and grid distortion were set to $p = 0.5$. For these three methods, only one operation, i.e. either elastic transform or grid distortion was allowed (*OneOf*). The degree of rotation was set to exactly 180, as the structures of carbon fibers used in our experiments always have a vertical alignment. *Elastic Transform* was performed with the parameters $\alpha = 10$, $\sigma = 10$, $alpha\_affine = 512 \cdot 0.05$ and $border\_mode = 4$. *Grid Distortion* was given the parameters $num\_steps = 2$ and $distort\_limit = 0.4$. In the respective datasets, the operations were applied online using the given parameters on every original input image. The U-Net model itself was slightly adapted from [42] to fit the dataset. The default size of the training images was 512×512, yet the default U-Net setting only accepts images of size 28×28. A ResNet-18 model is used as encoder by the U-Net. The architecture was adapted to fit the input size before applying the model. The training configuration of the U-Net is shown in Table 3.

Table 3. U-Net Configuration [27].

Parameter	Value
Learning rate	0.0001
Batch Size	10
Epochs	200
Loss Function	Binary Cross Entropy/Dice Loss

4.5 WGAN Configuration

The configuration for the WGAN training is shown in Table 4. As the WGAN produces non-binary image data as output, we applied a binarization stage to the final output images in order to get binary label masks.

Table 4. WGAN Configuration.

Parameter	Value
Learning rate	0.00005
Batch Size	64
Epochs	100,000

5 Evaluation

The metrics *accuracy*, *MCC* and $F_\beta - Score$ are less dependable for an objective evaluation of the classification model in our specific task. The proportion between background pixels (i.e. the non-defect pixels) and foreground pixels (i.e. the defect pixels) per image is thoroughly unbalanced, as the defects mostly consist of single fibers and therefore take much less space in the images. While *accuracy* returns the proportion of true results among all data points examined, *MCC* and $F_\beta - Score$ aim to balance out true and false positives and negatives of the binary classification result. In contrast, the *Jaccard index* or *Intersection over Union (IoU)* is used to measure the similarity of two sets, i.e. the similarity of the ground truth and the prediction. This property makes the *IoU* the most appropriate for the task at hand. Thus, we focus on the *IoU* metric for our experiments in order to allow an objective and problem related evaluation methodology. However, all relevant statisticals scores are reported in Table 5 for the sake of completeness.

Table 5. U-Net results from test runs on the datasets 1 through 6 for batch size 5, based on [27].

	PPV	TPR	IoU	ACC	MCC	F1	F2
Dataset 1	0.539169	0.586753	0.390778	0.985035	0.55487	0.561956	0.576576
Dataset 2	0.772803	0.718101	0.592925	0.991935	0.740872	0.744448	0.728413
Dataset 3	0.745926	0.721067	0.578888	0.991419	0.729034	0.733286	0.725905
Dataset 4	0.756767	0.705387	0.57502	0.991471	0.72631	0.730175	0.715098
Dataset 5	0.602418	0.585941	0.422541	0.990628	0.594065	0.543877	0.589383
Dataset 6	0.281570	0.433361	0.205801	0.980426	0.339847	0.341352	0.354881

5.1 Discussion of Results

For all of the six datasets the training was aborted after 200 epochs since it was clearly observable that the models had converged. As Fig. 9 suggests, both training accuracy and loss converge from epoch 100 onwards for the first four datasets. Training on dataset 1 was stopped at a loss rate of ~ 0.7, while for both datasets 3 and 4 the training ended at a loss rate of ~ 0.4. For dataset 2, model training reached an IoU of ~ 0.6 and ~ 0.5 for validation when the process was aborted. At the same time, the training loss ended at ~ 0.4 and reached a value of ~ 0.2 for validation. Furthermore, training on dataset 1 reached an IoU score of ~ 0.7 while dataset 3 and 4 achieved an IoU value of ~ 0.8 after 200 epochs.

Also, Fig. 10 shows that training on dataset 5 and 6 reached an IoU score of ~ 0.7 and ~ 0.5 for validation after 200 epochs. In both cases, the loss converged at ~ 0.2 for training and ~ 0.4 for validation. All training results for the four datasets are shown in Table 5.

The trained models were altogether evaluated on the test dataset. The model trained on dataset 1 reached an $accuracy$ of 0.985 and IoU of 0.391 on the test set, while the model trained with dataset 2 reached an $accuracy$ of 0.992 and an IoU of 0.593. Furthermore, the IoU for the model based on dataset 3 reached an IoU of 0.579 and an $accuracy$ of 0.991, while training with dataset 4 achieved a value of 0.575 for the IoU and 0.991 for the $accuracy$. In addition, dataset 5 resulted in an IoU of 0.423 and $accuracy$ of 0.991, while training on dataset 6 let the IoU drop to 0.206 and $accuracy$ decrease to 0.980. All metrics for the test runs were acquired from prediction on 25 randomly selected sample images as presented in Table 5.

Figure 8 shows a sample selection of defect images taken from the test set with red overlays representing the ROIs predicted by the U-Net model. It should be noted that training without any DA leads to more false positives which reminds of noise in the overlays as can be seen in Fig. 8b and 8c.

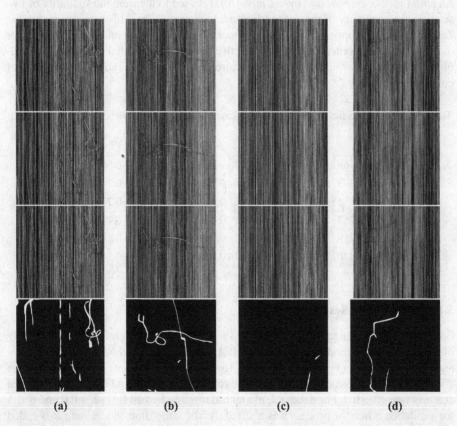

Fig. 8. Real carbon fiber defects from the test set with red overlay from U-Net segmentation for *dataset 1* (top row), *dataset 3* (second row), *dataset 5* (third row) and the *ground truth* (bottom row) [27]. (Color figure online)

For *dataset 1*, the loss rate drops heavily for 50 epochs before converging at a value of around 0.5 for the test set and just over 0.0 for the training set, as can be seen in Fig. 9. The IoU value increases for 50 epochs before slowing down and converging to an IoU of around 0.85 for the training set and around 0.4 for the test set after 125 epochs.

For *dataset 2* the loss rate drops considerably during the first 5 epochs, then decreases slowly but steady until it converges around 0.2 after 125 epochs during training. The loss on the test set shows a similar behaviour, except it converges around a value of 0.4. Again, the IoU value increases clearly within less than 5 epochs during the training process, then only slightly rises before converging around 0.6 after epoch 125. For the test set, the IoU value increases constantly between epoch 0 and 75, then converges around 0.4 as illustrated in the second row of Fig. 9.

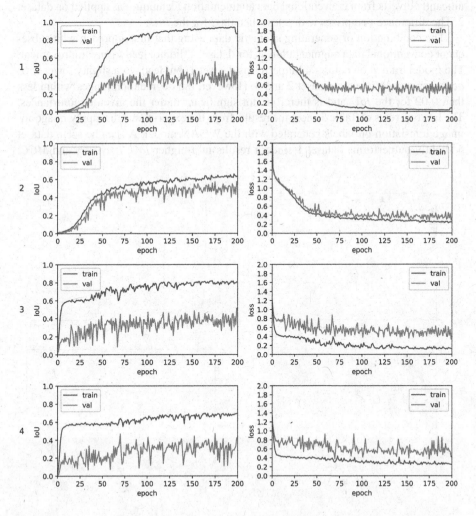

Fig. 9. IoU Score and Loss during U-Net training for dataset 1 through 4 from top to bottom row [27].

As can be seen in Fig. 10, the loss rate of dataset 5 drops heavily for 25 epochs, then quickly converges around 0.2 for the training set. The loss curve behaves the same way, although it oscillates more. The documented loss on dataset 6 shows the same characteristics, however its loss curve oscillates even more on the test set.

5.2 Assessment of Results and Insights

As can be seen, the U-Net model trained on dataset 3 significantly outperformed the model trained on dataset 1. This shows that our approach of mathematical defect modeling in combination with a *pix2pix* architecture could substantially improve the quality and diversity of the raw training set. When comparing the results of dataset 2 and dataset 3, it becomes apparent that the proposed approach is slightly worse, however not significantly diverts from conventional data augmentation techniques as applied on dataset 2. The difference comprises within less then 0.02 for the *IoU*.

The combination of generating synthetic data using that first approach with subsequent conventional data augmentation as for dataset 4 did not lead to any improvement. The model trained on dataset 4 outperforms dataset 1, but leads to a slighty lower *IoU*, *accuracy* and *MCC* than dataset 2 and 3. However, the degradation ranges within less then 0.02 for the *IoU* and is therefore not significant under the given circumstances. The last two rows of Table 5 show the results from the experiments that apply image-to-image translation on labels generated with the WGAN model. As can be seen, dataset 5 slightly outperforms dataset 1 since it results in a higher *IoU*, *accuracy* and *MCC*.

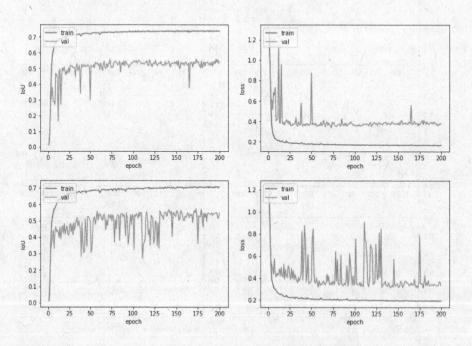

Fig. 10. IoU Score and Loss during U-Net training for dataset 5 and 6 from top to bottom row.

However, it clearly proves inferior to datasets 2 through 4 which means regular data augmentation as well as the problem-tailored label generator clearly performs better in the given scenario. For the sake of completeness and transparency, we also did an experiment in which we extended the data of *dataset 5* with conventional data augmentation, resulting in *dataset 6*. The outcome, though, indicates a clear deterioration all along the line. Every metric appears lower then for any other dataset (i. e. dataset 1–5 perform better). *Precision, IoU, MCC, $F_1 - Score$* and $F_2 - Score$ drop heavily - to values of 0.28, 0.21, 0.34, 0.34 and 0.35. The only metrics, that show a comparably 'light' degradation are *accuracy* (to 0.98) and *recall* (to 0.43) which is due to the fact that they weigh out background and foreground pixels. Since only $\sim 1\%$ of the pixels are associated with the region of interest, the *accuracy* and *recall* cannot provide high significance.

The experiment based on dataset 6 obviously suggests that online data augmentation does not add any value to WGAN generated labels and image-to-image translation. The numbers suggest that it even worsens the performance of the semantic segmentation model by a magnitude.

As the results from Table 5 and the samples depicted in Fig. 8 suggest, the pairs of synthetic images and labels of carbon fiber defects were successfully used to replace traditional data augmentation for semantic segmentation network training. With an IoU of 0.579, the first variant of our approach, i.e. the mathematical modeling of defects, performs comparably to U-Net training with regular data augmentation. Since the absolute difference between dataset 2 and 3 results in a value of 0.01, it appears negligible. The results show that synthesized training data helps to improve the detection quality of a U-Net segmentation model to a great extent. Moreover, the augmented dataset could

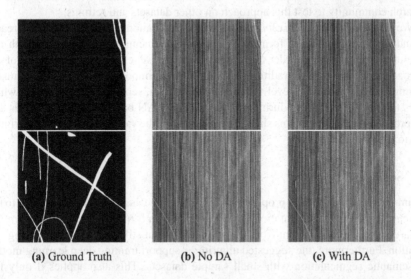

(a) Ground Truth   (b) No DA   (c) With DA

Fig. 11. Comparison of noticeable side effects in the classification of filaments based on the WGAN based approach with and without data augmentation and the corresponding ground truth.

be created based on few samples of only 300 images with an image size of 512×512 pixels in which the *ROI* on average only covers 1% of an image frame.

5.3 Visual Comparison and Critical Reflection

Since size of a ROI only take up a small portion of each image, even little variance in the detection results heavily affects the evaluation. This is the case if the classification of single pixels differs slightly from the ground truth. Adding a small margin to each filament e. g. largely affects metrics such as IoU or MCC and blurs the results of a proper analysis in this particular case. As can be seen in Fig. 8, the visual comparison of test results suggests that WGAN+Pix2Pix based semantic segmentation tends to return better segmentation than training a U-Net model without any data augmentation.

Actually, the ground truth sometimes reveals some room for misinterpretation by showing annotations that results in contradictory objective functions: as can be seen in Fig. 11, the ground truth differs substantially from all segmentation results; to the user this may not be obvious and visually the results range at a very similiar level, but little variations at this level cause the metrics to fluctuate heavily. Although this is intended to some extent, it must be considered closely upon interpretation. All in all, we consider this approach as promising and stress that future work should test it on other similiar or very different applications.

Moreover, the authors suggest to keep in mind that this study has only applied the workflow based on WGAN, Pix2Pix and trigonometric functions on images of a carbon fiber surface with its very special types of defects and anomalies. At this point, it cannot be ruled out that the workflow functions differently under other circumstances that are considerably distinct to the present use case. In fact, the authors greatly encourage the research community to test this approach on other datasets and settings.

We consider both our *pix2pix* based image generation approaches a more realistic and application-oriented form of DA. The experiments were conducted with and without traditional DA in order to evaluate the effectiveness and expandability of our approach. Excessive use of traditional DA might superimpose the 'real' data within the training set due to its low level form of manipulation, reproduction and reuse which raises the risk of overfitting during model training. GAN based data generation is also less prone to repetitive patterns since it tries to project the variation found in the original data to the synthetic data.

5.4 Summary

In summary, the approach as proposed in this article reveals a potential alternative to traditional, simple data augmentation. The generated data forms representation of images that are significantly different from the sample data but still resemble the training distribution. Furthermore, the suggested algorithms support training deep learning models for semantic segmentation with small sample datasets. This also applies if only few annotations are available.

In this article, two competing concepts were suggested and evaluated on industry data. While the first approach based on a trigonometric function represents a very problem-specific, hand-crafted solution, the second concept entirely uses adversarial and deep learning for model training. In both cases, the sample data needs to be anno-tated. Still, the number of sample data needed to train the models appears to be compa-rably low.

On the one hand, the approach based on WGAN, Pix2Pix and U-Net combines three different deep learning architectures and only requires parameter tuning to work for the given dataset. This workflow already offers a high level of automation since it reduces the effort for designing a defect detection system to providing a small sample set of annotated data. Of course, only a well-selected and sufficiently annotated test set allows for serious model validation and testing. On the other hand, the trigonomet-ric function is transparent, human-readable and stable. Its output can be visualized and tested against tolerance criteria. It also allows to be tuned by setting limiting parameters, such as window size, orientation or line thickness. Also, auditing and requirements test-ing can be easily performed. However, the mathematical function lacks of flexibility in terms of domain transfer. In order to design a mathematical model specific to the prob-lem, domain knowledge needs to be collected and translated into abstract dependencies. Thus, it qualifies for applications with a high need for transparency and stability, e. g. security in critical environments.

However the authors stress that they cannot evaluate the whole extent of GAN based DA, but encourage fellow researchers to explore the application of this approach to other fields and use cases. We expect benefits especially in the field of deep learning, industrial monitoring and neuroevolution.

6   Conclusion and Outlook

In this article, we presented image-to-image translation as a means for data augmenta-tion in the context of defect detection on textiles and carbon fiber in particular. There-fore, we discussed related GAN approaches and designed two variations of a novel concept for generating synthetic defects based on sparse labeled data using a *pix2pix* model.

Within our experiments on six different datasets we showed that a *pix2pix* based approach could substantially improve the pixel-based classification quality of U-Net models when using a problem-specific label generator. In general, the synthetic defects helped to augment the dataset so that segmentation quality improves on sparse data. However, the approach did not clearly outperform regular DA techniques. Still, domain-specific modelling of defect images allows to achieve similiar quality scores. Although WGAN proved inferior to competing techniques, it still helped to support semantic segmentation to a certain degree.

Furthermore, the approach can be used to train neural networks for semantic seg-mentation on comparably sparse data since GANs manage to generate realistic, yet artificial labels from few samples. Fuzzballs and misaligned fibers serve as a model for industrial camera based surface monitoring in the manufacturing process. The sug-gested approach has been tested for the given setting but is not limited to textiles. The example was selected in order to demonstrate how the approach behaves for complex detection tasks. The authors deem further experimentation necessary to evaluate the whole potential, i.e. explore and demonstrate the proposed workflow in other domains and for defects that are maybe even less complex. It should be mentioned, that by using the label generator with mathematical models, training data can always be created for a specific use case as for carbon fiber images. Conventional DA only applies very general image transformations without any reference to specific requirements in the application scenario. For that reason, we also suggested WGAN for a higher level of automation. In conclusion, we could reveal great potential for problem-tailored function models but could not verify the same for the WGAN based approach. The authors, however, suggest to conduct hyperparameter tuning on the GAN models for improvements in future work. The study suggests that semi-supervised training exhibits high relevance for defect detection in industrial applications. Furthermore, it could also be demonstrated that traditional DA did not substantially improve the pixel-based classification. Under the given circumstances the assumption can be made that GAN based augmentation already provides a well-balanced and diverse dataset so that conventional image transformation methods do not add any additional value.

Since the current research activities still focus on the adaptation of network models for semantic segmentation and object detection in different areas of industrial image data, we are going to explore the potential for a wide application of GAN training with very little supervision. For this purpose, future research will mainly address the design of networks which allow to design models that are trained with few or no data and still manage to effectively apply industrial monitoring solutions. This will result in problem-specific algorithm development in order to decrease efforts spent on defect simulation and shift it to domain independent methods. We deem the field of neuroevolution in combination with transfer and few-shot learning as promising for industrial applications. Future work will strive for a closer look to hyperparameter optimization in the context of deep learning. Therefore, we plan to extend our concepts with methods from the fields of *Evolutionary* and *Organic Computing* to equip our approach with self-configuring and self-learning properties. The application of evolutionary computation, i.e. genetic algorithms and co-evolution, constitutes another topic of our research agenda.

Acknowledgements. The authors would like to thank the Administration of Swabia and the Bavarian Ministry of Economic Affairs and Media, Energy and Technology for funding and support to conduct this research as part of the program *Competence Expansion of Fraunhofer IGCV formerly Fraunhofer Project Group for "Functional Lightweight Design" FIL of ICT*.

References

1. Arjovsky, M., Bottou, L.: Towards principled methods for training generative adversarial networks. arXiv preprint arXiv:1701.04862 (2017)
2. Arjovsky, M., Chintala, S., Bottou, L.: Wasserstein GAN. arXiv preprint arXiv:1701.07875 (2017)
3. Bowles, C., et al.: Gan augmentation: augmenting training data using generative adversarial networks. arXiv preprint arXiv:1810.10863 (2018)
4. Cavigelli, L., Hager, P., Benini, L.: CAS-CNN: a deep convolutional neural network for image compression artifact suppression. In: 2017 International Joint Conference on Neural Networks (IJCNN), pp. 752–759, May 2017

5. Choi, J., Kim, T., Kim, C.: Self-ensembling with GAN-based data augmentation for domain adaptation in semantic segmentation. In: Proceedings of the IEEE International Conference on Computer Vision, pp. 6830–6840 (2019)
6. Choi, Y., Choi, M., Kim, M., Ha, J.W., Kim, S., Choo, J.: StarGAN: unified generative adversarial networks for multi-domain image-to-image translation. In: Proceedings of the IEEE Conference on Computer Vision and Pattern Recognition, pp. 8789–8797 (2018)
7. Cui, Y.R., Liu, Q., Gao, C.Y., Su, Z.: FashionGAN: display your fashion design using conditional generative adversarial nets. Comput. Graph. Forum 37, 109–119 (2018)
8. Di Mattia, F., Galeone, P., De Simoni, M., Ghelfi, E.: A survey on GANs for anomaly detection. arXiv preprint arXiv:1906.11632 (2019)
9. Ferguson, M.K., Ronay, A., Lee, Y.T.T., Law, K.H.: Detection and segmentation of manufacturing defects with convolutional neural networks and transfer learning. Smart Sustain. Manuf Syst. 2 (2018)
10. Frid-Adar, M., Klang, E., Amitai, M., Goldberger, J., Greenspan, H.: Synthetic data augmentation using GAN for improved liver lesion classification. In: 2018 IEEE 15th International Symposium on Biomedical Imaging (ISBI 2018), pp. 289–293. IEEE (2018)
11. Geinitz, S., Margraf, A., Wedel, A., Witthus, S., Drechsler, K.: Detection of filament misalignment in carbon fiber production using a stereovision line scan camera system. In: Proceedings of 19th World Conference on Non-Destructive Testing (2016)
12. Geinitz, S., Wedel, A., Margraf, A.: Online detection and categorisation of defects along carbon fiber production using a high resolution, high width line scan vision system. In: Proceedings of the 17th European Conference on Composite Materials ECCM17, European Society for Composite Materials, Munich (2016)
13. Gulrajani, I., Ahmed, F., Arjovsky, M., Dumoulin, V., Courville, A.C.: Improved training of Wasserstein GANs. In: Advances in Neural Information Processing Systems, pp. 5767–5777 (2017)
14. Haselmann, M., Gruber, D.: Supervised machine learning based surface inspection by synthetizing artificial defects. In: 2017 16th IEEE International Conference on Machine Learning and Applications (ICMLA), pp. 390–395. IEEE (2017)
15. He, K., Zhang, X., Ren, S., Sun, J.: Deep residual learning for image recognition. In: Proceedings of the IEEE Conference on Computer Vision and Pattern Recognition, pp. 770–778 (2016)
16. Huang, S.-W., Lin, C.-T., Chen, S.-P., Wu, Y.-Y., Hsu, P.-H., Lai, S.-H.: AugGAN: cross domain adaptation with GAN-based data augmentation. In: Ferrari, V., Hebert, M., Sminchisescu, C., Weiss, Y. (eds.) ECCV 2018. LNCS, vol. 11213, pp. 731–744. Springer, Cham (2018). https://doi.org/10.1007/978-3-030-01240-3_44
17. Isola, P., Zhu, J.Y., Zhou, T., Efros, A.A.: Image-to-image translation with conditional adversarial networks. In: Proceedings of the IEEE Conference on Computer Vision and Pattern Recognition, pp. 1125–1134 (2017)
18. Krizhevsky, A., Sutskever, I., Hinton, G.E.: ImageNet classification with deep convolutional neural networks. In: Advances in Neural Information Processing Systems, pp. 1097–1105 (2012)
19. Liu, L., Zhang, H., Ji, Y., Wu, Q.M.J.: Toward AI fashion design: an attribute-GAN model for clothing match. Neurocomputing 341(2019). https://doi.org/10.1016/j.neucom.2019.03.011
20. Long, J., Shelhamer, E., Darrell, T.: Fully convolutional networks for semantic segmentation. In: Proceedings of the IEEE Conference on Computer Vision and Pattern Recognition, pp. 3431–3440, June 2015

21. Margraf., A., Hähner., J., Braml., P., Geinitz., S.: Towards self-adaptive defect classification in industrial monitoring. In: Proceedings of the 9th International Conference on Data Science, Technology and Applications - Volume 1: DATA, pp. 318–327. INSTICC, SciTePress (2020). https://doi.org/10.5220/0009893003180327

22. Margraf, A., Stein, A., Engstler, L., Geinitz, S., Hähner, J.: An evolutionary learning approach to self-configuring image pipelines in the context of carbon fiber fault detection. In: 2017 16th IEEE International Conference on Machine Learning and Applications (ICMLA). IEEE, December 2017

23. Mariani, G., Scheidegger, F., Istrate, R., Bekas, C., Malossi, C.: BAGAN: data augmentation with balancing GAN. arXiv preprint arXiv:1803.09655 (2018)

24. Masci, J., Meier, U., Ciresan, D., Schmidhuber, J., Fricout, G.: Steel defect classification with max-pooling convolutional neural networks. In: The 2012 International Joint Conference on Neural Networks (IJCNN), pp. 1–6. IEEE (2012)

25. McCann, M.T., Jin, K.H., Unser, M.: Convolutional neural networks for inverse problems in imaging: a review. IEEE Sig. Process. Mag. **34**(6), 85–95 (2017). https://doi.org/10.1109/msp.2017.2739299

26. Mertes., S., Baird., A., Schiller., D., Schuller., B., André., E.: An evolutionary-based generative approach for audio data augmentation. In: Proceedings of the 22nd International Workshop on Multimedia Signal Processing (MMSP). IEEE (2020)

27. Mertes., S., Margraf., A., Kommer., C., Geinitz., S., André., E.: Data augmentation for semantic segmentation in the context of carbon fiber defect detection using adversarial learning. In: Proceedings of the 1st International Conference on Deep Learning Theory and Applications - Volume 1: DeLTA, pp. 59–67. INSTICC, SciTePress (2020). https://doi.org/10.5220/0009823500590067

28. Pathak, D., Krahenbuhl, P., Donahue, J., Darrell, T., Efros, A.A.: Context encoders: feature learning by inpainting. In: Proceedings of the IEEE Conference on Computer Vision and Pattern Recognition, pp. 2536–2544 (2016)

29. Ren, S., He, K., Girshick, R., Sun, J.: Faster R-CNN: towards real-time object detection with region proposal networks. In: Cortes, C., Lawrence, N.D., Lee, D.D., Sugiyama, M., Garnett, R. (eds.) Advances in Neural Information Processing Systems, vol. 28, pp. 91–99. Curran Associates, Inc. (2015). http://papers.nips.cc/paper/5638-faster-r-cnn-towards-real-time-object-detection-with-region-proposal-networks.pdf

30. Rizki, M.M., Zmuda, M.A., Tamurino, L.A.: Evolving pattern recognition systems. IEEE Trans. Evol. Comput. **6**, 594–609 (2002)

31. Ronneberger, O., Fischer, P., Brox, T.: U-Net: convolutional networks for biomedical image segmentation. In: Navab, N., Hornegger, J., Wells, W.M., Frangi, A.F. (eds.) MICCAI 2015. LNCS, vol. 9351, pp. 234–241. Springer, Cham (2015). https://doi.org/10.1007/978-3-319-24574-4_28

32. Schlegl, T., Seeböck, P., Waldstein, S.M., Schmidt-Erfurth, U., Langs, G.: Unsupervised anomaly detection with generative adversarial networks to guide marker discovery. In: Niethammer, M., et al. (eds.) IPMI 2017. LNCS, vol. 10265, pp. 146–157. Springer, Cham (2017). https://doi.org/10.1007/978-3-319-59050-9_12

33. Shorten, C., Khoshgoftaar, T.M.: A survey on image data augmentation for deep learning. J. Big Data **6**(1), 60 (2019)

34. Shrivastava, A., Pfister, T., Tuzel, O., Susskind, J., Wang, W., Webb, R.: Learning from simulated and unsupervised images through adversarial training. In: Proceedings of the IEEE Conference on Computer Vision and Pattern Recognition, pp. 2107–2116 (2017)

35. Simonyan, K., Zisserman, A.: Very deep convolutional networks for large-scale image recognition. arXiv preprint arXiv:1409.1556 (2014)

36. Soukup, D., Huber-Mörk, R.: Convolutional neural networks for steel surface defect detection from photometric stereo images. In: Bebis, G., et al. (eds.) ISVC 2014. LNCS, vol. 8887, pp. 668–677. Springer, Cham (2014). https://doi.org/10.1007/978-3-319-14249-4_64

37. Staar, B., Lütjen, M., Freitag, M.: Anomaly detection with convolutional neural networks for industrial surface inspection. Procedia CIRP **79**, 484–489 (2019)

38. Stein, A., Margraf, A., Moroskow, J., Geinitz, S., Haehner, J.: Toward an organic computing approach to automated design of processing pipelines. In: 31th International Conference on Architecture of Computing Systems, VDE ARCS Workshop 2018 (2018)

39. Strumberger, I., Tuba, E., Bacanin, N., Jovanovic, R., Tuba, M.: Convolutional neural network architecture design by the tree growth algorithm framework. In: 2019 International Joint Conference on Neural Networks (IJCNN), pp. 1–8. IEEE (2019)

40. Szegedy, C., et al.: Going deeper with convolutions. In: Proceedings of the IEEE Conference on Computer Vision and Pattern Recognition, pp. 1–9 (2015)

41. Xie, S., Tu, Z.: Holistically-nested edge detection. In: Proceedings of the IEEE International Conference on Computer Vision, pp. 1395–1403 (2015)

42. Yakubovskiy, P.: Segmentation models (2019). https://github.com/qubvel/segmentation_models

43. Zhang, R., Isola, P., Efros, A.A.: Colorful image colorization. In: Leibe, B., Matas, J., Sebe, N., Welling, M. (eds.) ECCV 2016. LNCS, vol. 9907, pp. 649–666. Springer, Cham (2016). https://doi.org/10.1007/978-3-319-46487-9_40

Multi-stage Conditional GAN Architectures for Person-Image Generation

Sheela Raju Kurupathi[1,2]([✉]) [iD], Veeru Dumpala[1,2] [iD], and Didier Stricker[1,2]

[1] Department of Computer Science, Technical University of Kaiserslautern, Kaiserslautern, Germany
{sheela_raju.kurupathi,didier.stricker}@dfki.de
[2] Augmented Vision, German Research Center for Artificial Intelligence, Kaiserslautern, Germany

Abstract. Generating realistic human images has been of great value in recent times due to their varied application in Robotics, Computer Graphics, Movie Making, and Games. Advancements in Artificial Intelligence (AI) and Machine learning (ML) lead to the rapid growth of integrating every aspect into AI and ML. There are many deep learning models like Variational Auto Encoders (VAE), Stacked Hourglass networks and Generative Adversarial Networks (GANs) for solving human-image generation. However, it is still difficult for these models to generalize well to the real-world person-image generation task qualitatively. In this paper, we develop a multi-stage model based on Conditional GANs, which could synthesize the new image of the target person given the image of the person and the target (novel) pose desired in real-world scenarios. The model uses 2D keypoints to represent human poses. We propose a Multi-stage Person Generation (MPG) model, in which we have modified the Generator architecture of Pose Guided Person Image Generation (PG^2) resulting in two approaches. The first three-stage person generation approach has an additional generator integrated to base architecture and has trained the model end-to-end. The second two-stage person generation approach has a novel texture feature block in stage-II and has been trained incrementally to improve the generation of human-images qualitatively. The proposed two-stage MPG approach has generated promising results on Market-1501 Dataset. The claims are supported by benchmarking the proposed models with recent state-of-the-art models. We also show how the multi-stage conditional GAN architectures influence the process of human-image synthesis.

Keywords: GANs · Person-image generation · 2D keypoints

1 Introduction

Generating human images along with pose and clothing details has been one of the challenging tasks in the area of neural image synthesis. The development of

Supported by organization DFKI.

autonomous systems that could interact with humans in day-to-day life has been significantly increased. Humans can interpret the information from the surroundings without much difficulty, whereas developing the computational method with the same capabilities is a challenging task. It is not very easy since the method does not know about the varying backgrounds, articulated objects, occluded people, diverse lightning conditions, appearance variation due to complex body articulations and clothing. The person-image generation has been one of the most crucial tasks that have been discussed over the past few decades. The main problems while generating human images are the generation of the appearance details like clothing textures and representation of the human pose [1].

The human poses can be represented in various ways like 2D skeletons (stick figures), segmentation masks, 2D heat maps, 2D silhouettes, dense pose, and 3D pose skeletons described detailed in [1]. We need to choose the human pose representation that would generate human images with an accurate pose. Although such pose representations work well for human images, they fail in some cases of blurry and occluded human images. For generating the clothing textures, we can use Warping [2] and Clothing segmentation techniques. A wide range of deep learning models like PG^2 [3], Pix2pixHD [4], Deformable GANs [5] has been used for generating the human poses along with the clothing. The deep learning methods have surpassed the performance of traditional Computer Vision techniques for human image generation in terms of accuracy [6]. However, some of these models still suffer to generate human images with accurate pose and clothing due to many variations in the textures, appearance, shape, etc. The advantages of pose transfer, along with clothing, can be found in forgery detection and texture transfer applications. The generated human images can be used for training forgery detection algorithms with multiple realistic human images for the detection of forged images [7]. The use of Generative and Adversarial training would help to generate new realistic human images with varying background and poses that can be used in various Computer Vision applications for human activity recognition from RGB images. Few of the applications include detection of people behaviours in surveillance systems, self-driving cars, driver-assistance systems, virtual reality, gaming, robot assistance systems, etc.

In this paper, we address the challenges for human-image generation and develop GAN models for human-image generation conditioned on pose and appearance details. The data that we have used for training consists of challenging images with varying backgrounds along with occlusions, and the foreground may contain human with other deformable objects in hand. The deformable objects make the model difficult to learn and predict as they can be random. The main advantage of this research work could be found in generating the realistic images with varied poses of the same person that could be used in the fashion industry for reducing the burden to take images of the person in different poses, especially for online shopping portals. Availability of data to train a deep learning model is of utmost importance, and the data for human images is very scarce in both 2D and 3D domains. We can generate the human-images in rare poses by using the proposed model, which can be used as synthetic datasets for humans along with pose and clothing for future research purposes.

In order to generate the human-image with accurate pose and appearance details, the model needs to know the human body pose information. We represent the pose related information using the 2D keypoints representing the 2D coordinates for each joint in the image to avoid the expensive annotations for poses. We use the HumanPoseEstimator (HPE) [8] to estimate the 2D coordinates for all the human joints, and the number of keypoints we considered is 18. The model learns the positions of joints in the human body using these 18 key points. In our previous work [1], we have presented the single-stage Conditional GAN architecture [1] for human-image generation. In this work, we have used multi-stage Conditional GANs to deal with the problem of generating humans with pose and clothing details. We show how multi-stage GAN architecture has an effect on the refinement of the final human-image generation. However, training these multi-stage GANs requires high computational power and a sufficient amount of training data to achieve the desired results [9]. In this research paper, we do not make any assumptions about the objects, backgrounds, etc. and we do not use any explicit representations to denote the clothing information like semantic segmentation maps. We also do not give any information about the apparels like handbags, shoes, sandals, etc. by using any other explicit intermediate representations.

2   Related Work

Many deep learning applications use most commonly used deep learning architectures like AlexNet [10], ResNet [11], Google Net [12], etc. We use Generative Adversarial Network architecture to solve the problem related to human-image generation. Generative Adversarial Networks (GANs) [13] have shown substantial improvement in the neural image synthesis. GANs are aimed to generate novel data which has similar characteristics to that of real-world data and uses advanced methods to stabilize the training phase to improve the quality of generated images. The images can be translated from one domain to others when there is local alignment among them. Whereas in the case of human image generation, the input image has no alignment with the target image due to varying poses. Therefore, the image-to-image model [16], which is used for translating an image from one domain to another, cannot be used for the human-image generation when the background and foreground changes. The applications of GANs include neural image synthesis, image in-painting, super-resolution, semi-supervised learning, and more. They have been applied in large-scale image domains for solving different tasks like image manipulation, image generation and semantic manipulation. Even though there are quite a few attempts in recent years, most of the research on GANs is only restricted to natural images [4, 14, 15]. Deep generative models have been the most popular models in the context of image generation and manipulation in unsupervised and supervised settings. Conditional pose generation helps to synthesize a new image of a person, given a reference image of the person and a target pose. It has varied applications such as visualizing fashion outfits and generating animations. However, all these models suffer from the generation of human images with accurate pose and texture details.

Most of the pose-based models focus on pose detection and shape estimation of people from images. The most critical task would be to transfer the one pose of person to another person or the same person in a different pose. The subjects can have different deformable objects in the foreground and the background, thus making the model difficult to learn. It is also challenging to learn the pose and the clothing details simultaneously. The main idea of our proposed models is to guide the human-image generation process explicitly by an appropriate pose representation like 2D stick figures to enable direct control over the generation process. Si et al. [17] proposed a pose-based human image generation method that keeps the pose unchanged in novel viewpoints by dividing the foreground and background of the input person image. Lassner et al. [18] explored the generation of full-body people in clothing, by conditioning on the fine-grained body part segments. Omran et al. [19] proposed Neural Body Fitting (NBF) approach, which provides fine-grained control over all parts of the body fitting process. It is a hybrid architecture that integrates a statistical body model within a Convolutional Neural Network (CNN). From a semantic segmentation of the image or an RGB image, NBF directly predicts the parameters of the body model. Those parameters are then passed to Skinned Multi-Person Linear (SMPL) [20] to produce a 3D mesh.

Tang et al. [21] proposed a novel Multi-scale Conditional Generative Adversarial Networks (MsCGAN) approach conditioned on input person image and any target pose to generate a human-image in target pose. The appearance and texture of the generated human-image are consistent with the input image. MsCGAN is a multi-scale adversarial network consisting of two generators and two discriminators. One generator transforms the conditional person image into a coarse (blurry) image of the target pose. The other generator enhances the detailed quality of appearance of the generated person image through a local reinforcement network. Later, the outputs of the two generators are merged into a high-resolution image. The generated image that is down-sampled to multiple resolutions is given as the input to multi-scale discriminator networks. The proposed multi-scale generators and discriminators handle different levels of visual features. It helps to synthesize high-resolution human images with realistic appearance and texture details. Balakrishnan et al. [22] proposed a GAN network that decomposes the person image generation task into foreground and background generation and then combines them to form the final desired image. Zhu et al. [23] proposed a GAN for pose transfer, where the pose of a condition person is transferred to a target pose. The condition pose is transferred into a sequence of intermediate pose representations before transferring to the target pose. This transfer is carried out by a sequence of Pose-Attentional Transfer Blocks (PATBs). These PATBs blocks transfer certain regions they attend to, generating the human image progressively. Essner et al. [24] has combined both VAE [25] and U-Net [16] architectures to disentangle both appearance and pose information. However, appearance features are challenging to be encoded by a latent code representation with fixed length as it gives rise to several appearance misalignments.

Neverova et al. [26] adopted DensePose [2] as its pose representation. Dense-Pose representation contains more information about depth and body part segmentation to produce more clothing texture details. DensePose integrates the ideas from surface-based modeling with neural synthesis to perform accurate pose transfer. The expensive cost of acquiring the DensePose to represent the target pose hinders its applicability, in contrast to keypoint-based pose representation, which is cost-efficient. Ma et al. [3] proposed a more general approach to synthesize person images in any arbitrary (random) pose. Similarly, a conditioning image of the person and a target new pose defined by 18 joint locations is the input to our proposed models. The generation process is divided into two different stages as pose generation and texture refinement. The generation of appearance details like textures of the cloth makes building the models a very laborious task as the physical parameters of the cloth must be known in advance to obtain a realistic image. Horiuchi et al. [27] addressed the problem of human image generation by using deformable skip connections, self-attention, and spectral normalization in GAN. Several works [28–31] were inspired by the virtually try-on applications and made progress in changing the clothes of a given person while keeping the person pose by warping the clothes to fit the body topology of the given person. There are various modules like inpainting [26], warping [26], and latent sketch modules [18] which are used in different architectures for generating better pose and texture details.

Another line of approach [32] where multi-view supervision was used to train a two-stage system which can generate images from multiple views. Recently in the model [5], the authors showed that introducing a correspondence component in a GAN framework allows for substantially more accurate pose transfer but with the loss of texture details. Inpainting modules are used in the recent models to achieve a considerable level of detail in both image resolution and texture of cloth. There are 3D clothing models which automatically capture real clothing to estimate body shape, pose, and to generate new body shapes. However, the use of data-driven models for 3D clothing models to generate the real cloth wrinkles and textures would be challenging as we only have scarce 3D data for clothed people. Specifically, we have U-Net based architectures that are commonly used for pose-based person-image generation tasks [3,4,18,24,32]. Because of local information in input and the output images is not aligned, the skip connections are not well-suited to handle large spatial deformations. Contrary to this, the proposed models use deformable skip connections to deal with this misalignment problem and to share the local information from the encoder to the decoder. In our previous work [1], we proposed single-stage Conditional GAN models, to deal with accurate human pose generation along with clothing. We used Spectral Normalization (SN) in both the Discriminator D and Generator G along with adversarial hinge loss. We have done an ablation study to represent which components of the model are best for the human-image generation. We have made an observation that integrating the GANs with Spectral Normalization (SN) has stabilized the adversarial training and also generated highly realistic images. In this paper, we have proposed multi-stage conditional GAN models to deal with human pose generation along with clothing. We want to evaluate how the multi-stage conditional GAN models affect final human-image generation.

3 Classification of Pose Generation Methods

The problem of human image generation has been researched both in the context of 2D Image Processing and 3D Computer Vision for several years. Some of the human pose representations are used as an intermediate representation for training the models. As human images vary in their illumination, occlusions, background, and clothing, the model might find it very difficult to interpret all these challenges directly. So, we use intermediate representations as a simplification of the RGB image. Therefore, 2D or 3D keypoints play a vital role as they are used as input to the generative models for the human-image generation. Based on the task of generating human pose and shape, we can classify the models as model-based methods and learning-based methods.

3.1 Model-Based Methods

These methods use the parameters of the human body model to estimate the human pose and shape from the images. Generally, these methods require to fit a body model to the images. For the 3D human poses, the 2D pose representations are used as an intermediate step for the fitting procedure. SMPL model [20] estimates the 3D parameters of the body using the 2D keypoints extracted from the image. However, it is very ambiguous to estimate 3D information directly from 2D images [33,34].

3.2 Learning-Based Methods

On the other hand, learning-based methods predict the 2D or 3D keypoints from a single image of a person. Recently, CNN's have shown progress in estimating the 2D and 3D keypoints from the data taken from 2D and 3D datasets [33,34]. However, we have limited data of 3D keypoints for human images, and the variation in the data is significantly less. The images are also constrained to indoor environments, which make the models challenging to generalize well to real-world images. These models benefit from the usage of robust methods like regressing pose from the joint distances or by training any feed-forward neural network which could directly predict the 3D pose from 2D skeletons (stick figures).

In this paper, we have used 2D keypoints for representing the poses of condition and target images. We feed the target and condition poses along with condition image to GAN model so that the model has prior information about the pose. We use HPE [8], a learning-based method to estimate the poses for the condition and target images. This approach uses a non-parametric Part Affinity Fields (PAFs), a set of 2D vector fields which encode the orientation and location of limbs [8]. It uses a greedy bottom-up parsing approach and achieves high accuracy independent of the number of people present in the image. We observed that the use of a learning-based method for pose estimation had improved the results of our proposed models due to accurate pose information as input to the models.

4   Multi-stage Person Generation (MPG) Model

We use the generative models to transfer the person appearance from a given pose to the desired target pose simultaneously and retain most of the appearance details of the person. We address this problem by proposing a Multi-stage Person Generation (MPG) model, using the Pose Guided Person Image Generation (PG^2) [3] multi-stage architecture as the base model. In general, the multi-stage model performs better than single-stage methods for the human-image generation [35]. As at each stage, the architecture is composed of a generator that is capable of learning a spatial model for pose and appearance by communicating the refined uncertainty-preserving beliefs between stages implicitly [35]. This multi-stage model focuses on building multiple stages to improve the performance of human-image generation model. For a given input image of a person and a novel pose, the model synthesizes the person image in the novel pose. The model first extracts the pose of the person in a 2D skeleton using an HPE [8] model.

The three-stage person generation approach of the proposed model has a similar architecture of (PG^2) with an additional generator model, as shown in Fig. 2. The framework of the generator for this approach consists of 3 stages: pose integration, image refinement, and final image synthesis. In stage one, the condition image of the person and the target pose are given as input to a U-Net like generator network G_1 to generate the coarse image of the person in the novel pose as seen in Fig. 1. In stage two, the coarse result from stage one and the condition image is refined by training a U-Net like generator G_2 in an adversarial way. Finally, in stage three, the generator G_3 has U-Net like architecture that is trained to refine the image generated by the stage two and the condition image of the person to produce a final refined result. We have reported experimental results on both DeepFashion and Market-1501 datasets, which show that this approach generates images with less sharp details that are not so convincing.

In order to refine the image, even more, we proposed the second two-stage person generation approach that consists of two stages: pose integration and image refinement. Stage one is same as stage one of the three-stage approach, which would generate a coarse image of the person in a novel pose as seen from Fig. 3. In stage two, the condition image and the coarse result of stage one are concatenated and then given to a texture feature block. The texture feature block has convolutional layers inside and the output of this block is given to generator block G_2 that has U-Net like architecture (it is same as generator G_2 block in Fig. 2). The output from this G_2 block would be a difference map 1, as shown in Fig. 3. The output of the texture feature block is concatenated with difference map 1 after a Conv + BN + ReLU operation and given as input to another generator block G_2 which would give a refined difference map. This difference map is then added with a coarse image to obtain the refined result. In order to improve the refined result, we have trained the model incrementally. We have reported experimental results on Market-1501 dataset, which shows that this approach generates images with sharper details. Transferring the person from one pose into another is challenging for an end-to-end framework to generate pose and appearance details simultaneously.

4.1   Base Model

The main task of the model is to simultaneously transfer the appearance of a person from a given input pose to a desired (novel) pose by keeping the vital appearance details of the person. It uses explicit pose representation, in the format of keypoints to be able to model the diverse appearance of the human body. It consists of a two-stage approach to solve this task. In stage-I, a variant of U-Net is used to integrate the target pose along with the image of the person. As seen from Fig. 1, it generates a coarse result which captures the global structure of the human body in the desired target image of that person. Due to the use of masked L_1 loss, it generated blurry result at the end of stage-I. The main purpose of using this loss was to suppress any influence of background changes between the reference and target image. At stage-II, the generated coarse image of the person from the stage-I is further refined by using a variant of Deep Convolutional GAN (DCGAN) [36] model. The generator model in this stage learns to fill in more appearance details like clothing textures via adversarial training and tries to generate images with sharper details. In contrast to the regular use of GANs that learn to generate an image from scratch, the model is trained such that it generates a difference map. The difference map is calculated between the initial generated coarse image from stage-I and the target person image. This difference map makes the adversarial training to converge faster. The inclusion of masked L_1 loss is to regularize the training of the generator such that it would not generate any image with artifacts. Thus, it makes the model to focus on transferring the appearance details of the human body instead of the background information.

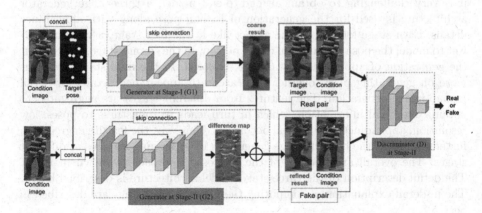

Fig. 1. Simple architecture of Pose Guided Person Image Generation (PG^2). Source: [3].

Architecture Details. The overall architecture of the base model is shown in Fig. 1. In stage-I, the encoder of generator G_1 consists of N residual blocks followed by one fully-connected layer. The value of N depends on the size of the input. Each residual block is composed of two convolution layers with a stride of 1 followed by one max-pooling layer with a stride of 2 except for the last residual block. In stage-II, the encoder of generator G_2 is fully convolutional, including $N - 2$ convolution blocks. Each convolution block consists of two convolution layers with a stride of 1 and one max-pooling layer with a stride of 2. The decoders in both G_1 and G_2 are symmetric in correspondence to the respective encoders. There are skip connections between respective decoders and encoders, as shown in Fig. 1. In both generators G_1 and G_2, there is no batch normalization or dropout applied. All the convolution layers consist of 3×3 kernels, and the number of kernels is increased linearly with each block. Then, it has a Rectified Linear Unit (ReLU) applied on each layer except the fully connected layer and the output convolution layer. The discriminator uses the architecture of DCGAN [36] except the size of the input convolution layer due to different image resolutions of the Market-1501 and DeepFashion datasets.

5   Multi-stage Person Generation (MPG) Approaches

5.1   1: Three-stage Person Generation Approach

The architecture is inspired by the divide and conquer strategy which divides the human-image generation into three stages for learning the entire human body structure along with clothing details similar to these models [37–39]. As it is very challenging to obtain an end-to-end model, a three-stage generator architecture is used for the generation of human pose along with appearance details. Each stage focuses on one aspect like in the first stage, a variant of U-Net to model the pose information of the person. In the second stage, more about the generation of appearance details are explored. In the final stage, we focus more on generating pose together with accurate appearance details for detail human-image generation. The features from different stages are aggregated to obtain contextual information, which in turn leads to robustness to poses, low resolution, appearance, and partial occlusions. This way, the feature aggregation further improves the human-image generation by leveraging features at different stages. The overall architecture of the proposed approach is shown in Fig. 2. The detail description of the three-stage model architecture is described below. The in-detail explanation of computing the losses at each stage are described in Sect. 5.3.

Stage-I: Pose Integration. In this stage, the conditioning person image I_A is concatenated with a target pose P_B of the person to generate a coarse or blurry image I_B. This initial coarse image captures the global structure of the human body as in the target image I_B. The state-of-the-art pose estimator [8] is used to obtain approximate poses of the human body. The pose estimator outputs the coordinates of 18 keypoints. Using these 2D keypoints as input to the model directly would require the model to learn the mapping between each keypoint to a position on the human body. Hence, the pose P_B is encoded as 18 heatmaps one for each keypoint. The gaussian heatmap is synthesized at the keypoint as the mean and standard deviation of 1 pixel. The generator at stage-I is a U-Net-like architecture [40] which consists of a convolutional autoencoder along with skip connections, as shown in Fig. 2.

The generator G_1 has stacked convolutional layers to integrate reference image I_A and target pose P_B from small local neighborhoods to larger neighborhoods. This stacking makes the appearance information of the human body to be easily integrated and transferred to neighboring body parts. Later, a fully connected layer is used so that the information between distant or far body parts can also be exchanged. In the decoder part of the generator G_1, there are set of stacked convolutional layers which are symmetrical to the convolutional layers of the encoder part. The generated image in the stage-I is denoted as $\hat{I}_{B1}$ and the skip connections between encoder and decoder of the U-Net architecture help to propagate information in the image directly from input to output. Use of residual blocks as essential component improves the performance of human-image generation. The residual block of stage-I, in contrast to the residual block [11] has only two consecutive convolutional and ReLU layers. In order to compare the generation correctness of coarse image $\hat{I}_{B1}$, with the target image I_B, L_1 distance is used as the generation loss in stage-I. The generated image is blurry because the L_1 loss is the average of all possible cases [16]. The generator G_1 captures the global structural information specified by the target pose, as shown in Fig. 2. It also captures other low-frequency information like the color of clothes. The high-frequency information like body appearance will be refined at the stage-II by adversarial training.

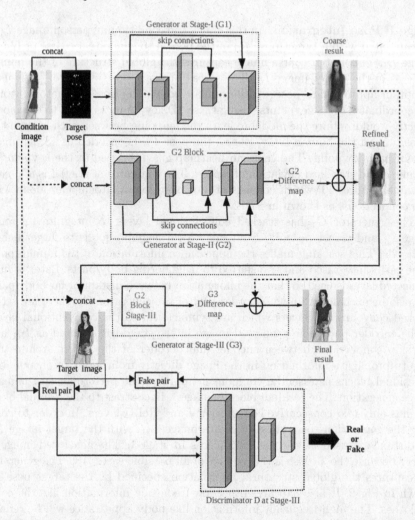

Fig. 2. Proposed architecture of the three-stage person generation approach of MPG model.

Stage-II: Image Refinement. The generator G_1 at stage-I, has synthesized the image of a person which is blurry but resembles in pose and basic color to the target image. The stage-II focuses on the generation of human-image with filling in the missing finer details in the generated image of stage-I. In stage-II, a variant of conditional DCGAN [36] is used as a generator G_2 and then conditioned on the generated output from stage-I. The conditional DCGAN [36] in stage-II is referred to as G_2 block, which is also used in stage-III as shown in Fig. 2. Since the initial result generated from stage-I and the target image are almost similar in their structure, the generator G_2 at stage-II is used to generate an appearance difference map. This map helps to make the initial result closer to the target image. The difference map is computed using a similar U-Net architecture as of stage-I. However, the inputs at stage-II are the initial generated result $\hat{I}_{B1}$ from stage-I and condition image I_A. We differ from U-Net architecture in terms of removing the fully-connected layer as it helps to preserve more details about body appearance from the input. It is due to the reason that a fully-connected layer compresses much information which is in the input. The difference map is then added to the coarse result of stage-I to generate the refined result $\hat{I}_{B2}$. The main purpose of difference maps is that it speeds up the convergence of the model during training. The model focuses on learning the missing appearance details like clothing, facial features, etc. instead of generating the target image from scratch. In particular, the training already starts from a reasonable result. The overall architecture of G_2 is shown in Fig. 2.

The discriminator D is used to distinguish between the real or ground truth and fake or synthesized images. As the generator G_2 takes the condition image I_A as input which is the natural image of the person, it may be misleading for the generator G_2 to directly output the input image I_A instead of refining the coarse result of the stage-I. Therefore, the output of G_2 is paired with the input condition image I_A, such that the discriminator can easily detect the extent of fakeness between the two pairs $(\hat{I}_{B2}, I_A)$ and (I_B, I_A) as shown in Fig. 2. The inclusion of pairwise input to the discriminator D encourages it to learn the exact distinction between $\hat{I}_{B2}$ and I_B.

Stage-III: Final Image Synthesis. The generated result at stage-II still lacks in few more appearance details when compared to the target image. We use generator G_3 at stage-III to further refine the results of stage-II making the appearance details sharper. With an additional generator G_3, we believe that the model would be capable of learning a spatial model for both pose and appearance by communicating between stages implicitly. In stage-III, generator G_3 is a variant of conditional DCGAN [36] (G_2 Block from the Fig. 2) which is conditioned on generated refined result of stage-II $\hat{I}_{B2}$ and the input condition image I_A. The generator G_3 in stage-III is trained to generate an appearance difference map so that the initial generated coarse image looks even closer to the target image I_B. The difference map is computed by using similar architecture as of stage-II with $\hat{I}_{B2}$ and the input condition image I_A as inputs. The difference map is then added with the refined result of stage-II $\hat{I}_{B2}$ to generate the final image with sharper appearance details. The convergence of model during the training speeds up by the difference maps. The discriminator D distinguishes between the ground truth and synthesized images. The generator G_3 takes the condition image I_A as input in contrast to the traditional GANs that taken random noise as input. The output of G_3 is paired with the input condition image I_A to enable the discriminator to easily detect the extent of fakeness between the two pairs $(\hat{I}_{B3}, I_A)$ and (I_B, I_A), as shown in Fig. 2. The idea of pairwise input to the discriminator D helps it to learn the exact distinction between $\hat{I}_{B3}$ and I_B.

5.2   2: Two-stage Person Generation Approach

As we already discussed that results generated by the three-stage approach are not so convincing, we proposed another approach in order to improve the generated result of approach 1. The three-stage approach takes only the outputs from the previous stage as the input for the current stage while our two-stage approach takes both the modified inputs and outputs from the previous stage for the current stage. We use two-stage generator architecture for the generation of human pose along with clothing details. The purpose of each stage is to focus on one of each aspects pose and appearance. In the stage-I, a variant of U-Net is used to model the pose information of the person same as in approach 1. In the second stage, we emphasize more on generating pose together with accurate appearance details for detail human-image generation. The stage-II is different in its training procedure and the architecture. The overall architecture of the two-stage approach is shown in Fig. 3. The detail description of architecture for the two-stage person generation model is described below. The in-detail explanation of computing the losses at each stage are described in Sect. 5.3.

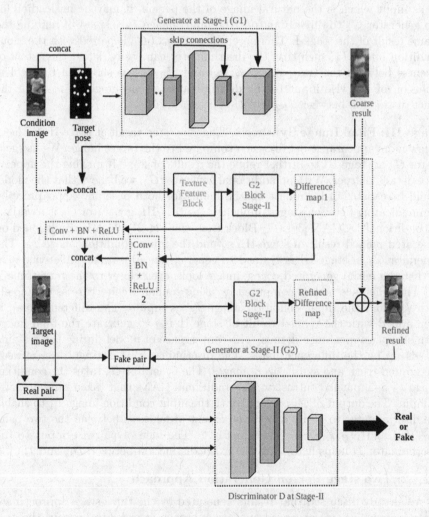

Fig. 3. Proposed architecture of the two-stage person generation approach of MPG model.

Stage-I: Pose Integration. In stage-I, the condition image of the person I_A is concatenated with a target pose P_B of the person to generate a coarse image I_B. The initial coarse image captures only the global structure of the human body as in the target image I_B, as shown in Fig. 3. The architecture for the stage-I is same as stage-I of the three-stage approach. It captures other low-frequency information like the color of clothes. The high-frequency information like appearance details will be refined at the stage-II by adversarial training.

Stage-II: Image Refinement. Since the initial result of stage-I $\hat{I}_{B1}$ is blurry, we want to refine the image further using stage-II. Stage-II focuses not only on the generation of a person with details missing in the initial result but refining the coarse result to a sharp image. In stage-II, the condition image I_A and the coarse image of stage-I $\hat{I}_{B1}$ are concatenated. The concatenated image is given as input to the texture feature block. This block is composed of two convolution layers with a stride of 1, where after the first convolution layer we have Batch Normalization (BN), and then ReLU activation applied over it. The main purpose of this texture feature block is that it is used to compute the features that carry information about the global structure. These features are aggregated with the features of difference map 1 to obtain both local detailed information and global context information for robust human image generation. This global information can highlight the important local parts further refining the human-image generation. The output of the texture feature block is then given to G_2 block, which would then generate a difference map 1, as shown in Fig. 3.

This map helps to make the initial result closer to the target image. The difference map is computed using a similar U-Net architecture as of stage-I. However, the inputs at stage-II are the initial coarse result $\hat{I}_{B1}$ from stage-I and condition image I_A that are passed to texture feature block after concatenating them. The only difference lies in the fully-connected layer that is removed from the U-Net architecture as it helps to preserve more details about body appearance from the input. A fully-connected layer compresses much information which is in the input due to which it is removed. The output of this texture feature block and the difference map 1 generated are concatenated after applying a Conv + BN + ReLU operation on each of them. The convolution layer applied on the output of texture feature block and difference map 1 consists of 3×3 kernels with a stride of 1. The concatenated output is then fed to the G_2 block which would generate a refined difference map. The refined difference map obtained would be sharper than the difference map 1, which indicates that the model has learned the sharper appearance and global structure of the human body. Now the refined difference map is added to the coarse image of stage-I to generate the refined result. The main purpose of difference maps is that it speeds up the convergence of the model during training. The model focuses on learning the missing appearance details like clothing and facial features instead of generating the target image from scratch.

We train this two-stage person generation approach incrementally, which means after training G_1, G_2 is trained. While training G_2, we incrementally update the generator G_1 for every 1000 iterations of training G_2. This way we can fine-tune both generators G_1 and G_2 so that the learning could be even more efficient.

5.3 Loss Functions

For 2-stage and 3-stage approaches, in stage-I, the L_1 distance is used in order to compare the coarse generation $\hat{I}_{B1}$ with the target image I_B. Since the input for the stage-I was a condition image I_A and a target pose P_B, it is very challenging for the model to generate the background of the target image if it has a different background from the condition image I_A. Therefore, in order to alleviate the influence of changes in the background on the final image generation, the L_1 loss is modified by adding the term pose mask M_B such that it adds more weight to the human body than the background. The pose mask loss is computed as below:

$$\mathcal{L}_{G1} = ||(G_1(I_A, P_B) - I_B) \odot (1 + M_B)||_1 \tag{1}$$

where $\odot$ denotes the pixels-wise multiplication operator. The pose mask M_B is set to 1 (white) for foreground and 0 (black) for the background, as detailed in [3]. It is computed by connecting all the human body parts and then applying a set of morphological operations like dilation and erosion [41] such that it can approximately cover the whole human body in the target image. However, the output of stage-I is blurry, and this needs to be refined for missing details in the later stages.

As in traditional GANs, the discriminator distinguishes between fake images which are generated from random noise and the real ground truth images. Whereas, in the conditional GAN, G_2 takes the condition image I_A as input instead of random noise. The output of G_2 is combined with the input condition image I_A so that the discriminator D in stage-II has $(I_A, \hat{I}_{B2}, I_B)$ triplet as input. Then, we have two pairs $(\hat{I}_{B2}, I_A)$ and (I_B, I_A) for the discriminator to distinguish synthesized image $\hat{I}_{B2}$ from the target image I_B. The loss functions for the discriminator D and the generator G_2 at stage-II for both the approaches are as follows:

$$\mathcal{L}_{adv}^D = \mathcal{L}_{bce}(D(I_A, P_B), 1) + \mathcal{L}_{bce}(D(I_A, G_2(I_A, \hat{I}_{B1})), 0) \tag{2}$$

$$\mathcal{L}_{adv}^G = \mathcal{L}_{bce}(D(I_A, G_2(I_A, \hat{I}_{B1})), 1) \tag{3}$$

where $\mathcal{L}_{bce}$ represents Binary Cross-Entropy (BCE) loss. Using both the adversarial loss with a loss minimizing $\mathcal{L}_p$ distance would help to regularize the image generation process [16,42]. Therefore, the masked $\mathcal{L}_1$ loss is used in the stage-I such that it gives more attention to the appearance details of target person than its background details. The loss function for generator G_2 for both the approaches is calculated as below:

$$\mathcal{L}_{G2} = \mathcal{L}_{adv}^G + \lambda ||(G_2(I_A, \hat{I}_{B1}) - I_B) \odot (1 + M_B)||_1 \tag{4}$$

where λ denotes the weight of $\mathscr{L}_1$ loss which controls how close the generated image looks like the target image at low frequencies. If the value of λ is small, then the adversarial loss $\mathscr{L}_{adv}^G$ would dominate the training. It would, in turn, lead to the generation of artifacts. On the other hand, if the value of λ is large, then the generator with $\mathscr{L}_1$ loss would dominate during training. It makes the model to generate blurry results. During the training of DCGAN, both the discriminator D and generator G_2 are alternatively optimized. The generator G_2 takes the output of stage-I $\hat{I}_{B1}$ and the condition image I_A as input and aims to refine the image to fool the discriminator D. Then, the discriminator D learns to classify the pair of condition image I_A and the generated image $\hat{I}_{B2}$ as fake while the other pair consisting of target image as real. The refined result at stage-II still lacks in few details like appearance for the three-stage approach, which lead to the addition of one more stage for final refinement. In the case of the two-stage approach, the refined result at stage-II is sharper due to better pose and appearance details.

For the three-stage person generation approach, in the stage-III, generator G_3 takes the condition image I_A as input along with the generated refined result of stage-II $\hat{I}_{B2}$. Similar to the stage-II, even in this stage $(I_A, \hat{I}_{B3}, I_B)$ triplet are given as input to the discriminator D. Hence, the two pairs $(\hat{I}_{B3}, I_A)$ and (I_B, I_A) help the discriminator to distinguish final synthesized image $\hat{I}_{B3}$ from the desired target image I_B. The respective loss functions for the discriminator D and the generator G_3 at stage-III for three-stage person generation approach are given below:

$$\mathscr{L}_{adv}^D = \mathscr{L}_{bce}(D(I_A, P_B), 1) + \mathscr{L}_{bce}(D(I_A, G_3(I_A, \hat{I}_{B2})), 0) \qquad (5)$$

$$\mathscr{L}_{adv}^G = \mathscr{L}_{bce}(D(I_A, G_3(I_A, \hat{I}_{B2})), 1) \qquad (6)$$

In the above equations, the term $\mathscr{L}_{bce}$ represents Binary Cross-Entropy (BCE) loss. A combination of the masked $\mathscr{L}_1$ loss and the adversarial loss $\mathscr{L}_{adv}^G$ are used at stage-II, it would help to regularize the image generation process. The loss function for the generator G_3 for three-stage person generation approach is computed as below:

$$\mathscr{L}_{G3} = \mathscr{L}_{adv}^G + \lambda ||(G_3(I_A, \hat{I}_{B2}) - I_B) \odot (1 + M_B)||_1 \qquad (7)$$

where λ represents the weight of $\mathscr{L}_1$ loss, and it controls the generation of image and how close it looks like the target image. During the training of conditional GANs, both the discriminator D and generator G_3 are optimized alternatively. The generator G_3 takes the output of stage-II $\hat{I}_{B2}$ and the condition image I_A as input and aims to further refine the image to fake the discriminator D. Thus, the discriminator D learns to classify the pair of condition image I_A and the generated image $\hat{I}_{B3}$ as fake while the other pair consisting of target image I_B as real. The generated image at the stage-III was sharper than the stage-II of three-stage person generation approach with refined appearance details but not so convincing than the refined results of stage-II of two-stage person generation approach.

6   Experimental Results

In this section, the evaluation results of the proposed models are reported and compared with other state-of-the-art methods for the human-image generation. We present both quantitative and qualitative evaluation results on publicly available Market-1501 and DeepFashion datasets. We described in detail about the evaluation metrics, optimizers used along with the implementation details for the proposed models.

The proposed Multi-stage Person Generation (MPG) model has two approaches of which the three-stage person generation approach was evaluated on publicly available Market-1501 [43] and DeepFashion [44] datasets that consist of diverse poses of the humans. In contrast, the two-stage person generation approach was evaluated only on Market-1501 [43] dataset. The approaches were trained with RMSprop and Adam optimizer, and we reported the results obtained by Adam optimizer as it could converge faster. We have compared our approaches with the most recent related work for human-image generation task [3,5].

6.1   Datasets

We used the most commonly available public datasets for humans like Market-1501 [43] and DeepFashion [44] datasets. **Person Re-identification dataset: Market-1501** This dataset contains images of 1,501 persons captured from 6 different surveillance cameras totaling to 32,668 images. It contains low-resolution images of size 128×64 with the high diversity in illuminations, poses, and backgrounds. **The DeepFashion dataset (In-shop Clothes Retrieval Benchmark)** comprises of 52,712 clothes images, leading to 200,000 pairs of same clothes with two different poses and scales of the persons wearing these clothes. These images have a resolution of 256×256 pixels, and the images are less noisy compared to the Market-1501 dataset. The dataset contains images with no background, where the clothing patterns can contain text, graphical logos, etc.

6.2   Evaluation Metrics

For benchmarking the performance of proposed models, widely-used evaluation metrics for human image generation like Structural Similarity [45] (SSIM), Inception Score (IS) [46] (based on the entropy of classification neurons), m-SSIM (masked SSIM), m-IS (masked IS) are calculated. Though these measures are widely accepted, we like to show the qualitative results of the generator with adversarial training that are visually appealing compared to present state-of-the-art approaches. The masked scores are obtained by masking out the background of the images and feeding it to the generator G, making it difficult to estimate how the background looks. Another metric DS (Detection Score) that is based on the detection outcome of the object detector SSD (Single Shot multibox Detector) [47] is widely used. During testing, person-class detection scores of SSD are computed on each generated image and averaging the SSD score of each generated image; the final DS is obtained. DS gives the confidence that a person is present in the image but not for evaluating the pose and realism of appearance details like clothing textures. Therefore, we did not consider this metric as part of our quantitative evaluation. All the metrics computed using the ground truth images (Real data) for the test set are reported in the tables which can be seen as the maximum value of the generated result but not in a strict case.

6.3 Implementation Details

The Adam [49] optimizer with learning rate $lr = 2 \times 10^{-5}$, hyperparameters $\beta_1 = 0.5$ and $\beta_2 = 0.999$ is used for both Market-1501 [43] and DeepFashion [44] datasets. For the DeepFashion dataset, the value of the number of convolution blocks is taken as $N = 6$. The three-stage person generation approach is trained with a mini-batch of size 8 for 30k, 20k, and 40k iterations at stage-I, stage-II and stage-III respectively. For the Market-1501 dataset, the number of convolution blocks is initialized as $N = 5$. The three-stage approach is trained with a mini-batch of size 16 for 30k, 20k, and 40k iterations at stage-I, stage-II and stage-III respectively.

For the two-stage person generation approach, we used Adam [49] optimizer with learning rate $lr = 2 \times 10^{-5}$, hyperparameters $\beta_1 = 0.5$ and $\beta_2 = 0.999$ for Market-1501 [43] dataset. We trained the two-stage approach with the number of convolution blocks as $N = 5$ with batch size 16 for 30k, 24k iterations at stage-I and stage-II respectively. For data augmentation, only left-right flip is applied for both Market-1501 [43] and DeepFashion [44] datasets.

6.4 Quantitative Results

Table 1. Quantitative results: Results of the proposed three-stage person generation approach on Market-1501 and DeepFashion datasets.

Model	DeepFashion			
	SSIM	IS	m-SSIM	m-IS
Three-stage approach	0.623	2.509	0.922	3.201
Model	Market-1501			
	SSIM	IS	m-SSIM	m-IS
Three-stage approach	0.091	3.143	0.699	3.294

Results of Three-stage Approach of the MPG Model: Market-1501 and DeepFashion. The quantitative results of the proposed three-stage approach are reported in Table 1 on DeepFashion and Market-1501 datasets. The IS scores obtained by the three-stage model on DeepFashion dataset are better than Market-1501 dataset as observed from Table 1. The SSIM score on Market-1501 dataset is 0.091, which is less than the SSIM score on DeepFashion dataset, i.e., 0.623.

Comparison of Two-stage and Three-stage Approaches of the MPG Model: Market-1501. The quantitative results of the proposed three-stage and two-stage approaches are reported in Table 2 on Market-1501 dataset. The scores on Market-1501 dataset has been improved rapidly by using our two-stage approach than the three-stage approach, as shown in Table 2. The bold measures in Table 2 represent the highest scores.

Table 2. Quantitative results: Results of the proposed three-stage and two-stage person generation approaches on Market-1501 dataset. For all the measures, higher is better.

	Market-1501			
Model	SSIM	IS	m-SSIM	m-IS
Three-stage approach	0.091	3.143	0.699	3.294
Two-stage approach	**0.232**	**4.405**	**0.775**	**3.610**
Real-Data	1.00	3.86	1.00	3.36

Comparison of MPG Model Approaches with Other Recent State-of-the-Art Models: Market-1501. Table 3 compares all proposed MPG model approaches with the benchmark models [3,5] on Market-1501 dataset. The bold scores from each column in Table 3 denote the highest scores. It shows that SSIM and m-SSIM scores obtained by variant-3 (W) are higher with a value of 0.293 and 0.805, respectively. However, m-SSIM scores for benchmark model [5] and our SN + H + W variant are observed to be same. The scores IS and m-IS are higher for our two-stage approach with 4.405 and 3.610 values, respectively. It

Table 3. Quantitative results: Comparison of results of the proposed MPG model approaches with the benchmark models [3,5] on Market-1501 dataset.

	Market-1501			
Model	SSIM	IS	m-SSIM	m-IS
Ma et al. [3]	0.253	3.460	0.792	3.435
Siarohin et al. [5]	0.290	3.185	**0.805**	3.502
Ours Three-stage approach	0.091	3.143	0.699	3.294
Ours Two-stage approach	0.232	**4.405**	0.775	**3.610**
SN [1]	0.280	3.300	0.797	3.528
SN + H [1]	0.291	3.239	0.804	3.592
W [1]	**0.293**	3.354	**0.805**	3.540
SN + H + W [1]	0.291	3.192	**0.805**	3.551
Real-Data	1.00	3.86	1.00	3.36

shows that the use of multi-stage architecture, has improved the inception scores as at each stage, the model is composed of a generator that is capable of learning a spatial model for both pose and appearance details. The features at each stage are cascaded to the next stage that refines them to obtain a better-generated image. However, just replicating the generators at each stage like our three-stage approach would not generate promising results. We need to cascade the features from the previous stage to the next stage along with different generator architecture like our two-stage approach, which would yield a better person generation. We believe that even incremental training procedure for fine-tuning the model would be beneficial in achieving promising results.

6.5   Qualitative Results

Results of Three-stage Approach of MPG Model: Market-1501 and DeepFashion. Figures 4 and 5 represent the qualitative results of the proposed three-stage approach on DeepFashion and Market-1501 datasets, respectively. The generated images of DeepFashion are not significantly sharper in appearance details. These images show the sharper generation of facial features. In Fig. 5, the generated images from the three-stage approach on Market-1501 dataset lack in the appearance details due to the noise in the condition image.

Results of Two-stage Approach of MPG Model: Market-1501. Figure 6 shows the qualitative results of the proposed two-stage approach on Market-1501 dataset. The generated images of Market-1501 are little sharp in their global structure and appearance details. However, the generated images still lack in facial features and clothing textures. It shows that the generation of a sharp structure is due to the high value of IS and m-IS scores, but the appearance details are not still robust to noise.

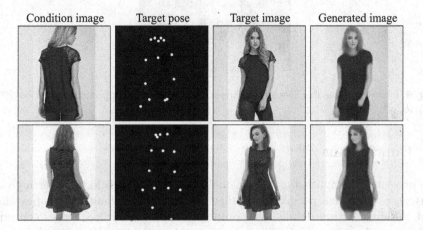

| Condition image | Target pose | Target image | Generated image |

Fig. 4. Qualitative results: Results of the proposed three-stage approach on DeepFash-ion dataset.

Condition   Target   Target   Generated
image   pose   image   image

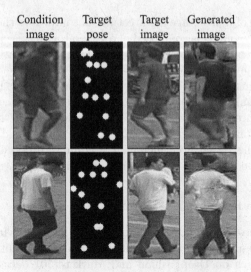

Fig. 5. Qualitative results: Results of the proposed three-stage approach on Market-1501 dataset.

Condition   Target   Generated
image   image   image

Fig. 6. Qualitative results: Results of the proposed two-stage approach on Market-1501 dataset.

7   Conclusions

We presented the multi-stage conditional generative models for transferring various poses from one person to another (same person) along with appearance and clothing texture details. The proposed MPG model approaches converge faster and also generalize well to never seen test data without any data augmentation. We showed how our proposed two-stage model outperforms recent state-of-the-art models for person image generation with pose and clothing details by benchmarking on publicly available Market-1501 dataset. The use of multi-stage architectures for tackling the problem of human-image generation has proved to improve the generation process. It can be seen from the significant increase in inception scores for our two-stage approach. The use of a novel texture feature block in our two-stage approach showed improvement in qualitative results when compared to the three-stage approach. The use of multi-stage conditional generative modeling for person image generation in a supervised setting could provide efficient human-image generation preserving pose and appearance details.

8   Future Scope of Research

There are many improvised models for GANs which are designed to tackle the problem of human image generation along with pose and clothing. Different models focus on different pose representations, but the combination of a good pose representation along with objective function would improve the person-image generation. In future work, we would like to focus on using other pose representations like 3D keypoints along with Spectral Normalization (SN) and hinge loss used in [1], which we believe would make better improvements in this area of research. Also, self-attention [52] modules could be integrated into the multi-stage Conditional GAN architectures along with different pose representations for the person-image generation. These attention modules would attend to all the important features of the person to generate accurate appearance details with high quality. The task of generating people in clothing along with the pose and high resolution is one of the long-standing goals in the area of Computer Vision. The representation of clothing would require domain knowledge about the clothing textures. The clothing models are complicated to construct since we need to have an idea about the physical parameters of the cloth, which can drive the model to generate realistic results. It is very laborious to model the interactions between the clothing and body. Thus, the complexity and cost mark the limiting case for diverse clothing simulation applications.

References

1. Kurupathi, S., Murthy, P., Stricker, D.: Generation of Human Images with Clothing using Advanced Conditional Generative Adversarial Networks. In: Proceedings of the 1st International Conference on Deep Learning Theory and Applications, pp. 30–41. SciTePress, France (2020). https://doi.org/10.5220/0009832200300041
2. Alp Güler, R., Neverova, N., Kokkinos, I.: Densepose: Dense human poseestimation in the wild. In: Proceedings of the IEEE Conference on Computer Vision andPattern Recognition, pp. 7297–7306 (2018)
3. Ma, L., Jia, X., Sun, Q., Schiele, B., Tuytelaars, T., Van Gool, L.: Pose guided person image generation. In: Advances in Neural Information Processing Systems, pp. 406–416 (2017)

4. Wang, T.-C., Liu, M.-Y., Zhu, J.-Y., Tao, A., Kautz, J., Catanzaro, B.: High-resolution image synthesis and semantic manipulation with conditional gans. In: Proceedings of the IEEE Conference on Computer Vision and Pattern Recognition, pp. 8798–8807 (2018)
5. Siarohin, A., Sangineto, E., Lathuilière, S., Sebe, N.: Deformable gansfor pose-based human image generation. In: Proceedings of the IEEE Conference on Computer Vision and Pattern Recognition, pp. 3408–3416 (2018)
6. Walsh, J., et al.: Deep learning vs. traditionalcomputer vision (2019)
7. Rössler, A., Cozzolino, D., Verdoliva, L., Riess, C., Thies, J., Nießner, M.: Face-forensics: A large-scale video dataset for forgery detection in human faces. arXivpreprint arXiv:1803.09179 (2018)
8. Cao, Z., Simon, T., Wei, S.-E., Sheikh, Y.: Realtime multi-person 2d poseestimation using part affinity fields. In: Proceedings of the IEEE Conference on Computer Vision and Pattern Recognition, pp. 7291–7299 (2017)
9. Stewart, M.: Advanced Topics in Gen-erativeAdversarialNetworks(GANs). https://towardsdatascience.com/comprehensive-introduction-to-turing-learning-and-gans-part-2-fd8e4a70775 (2019) (Accessed May 8 2019)
10. Krizhevsky, A., Sutskever, I., Hinton, G.E.: Imagenet classification withdeep convolutional neural networks. In: Advances in Neural Information Processing Systems, pp. 1097–1105 (2012)
11. He, K., Zhang, X., Ren, S., and Sun, J.: Deep residual learning for imagerecognition. In: Proceedings of the IEEE Conference on Computer Vision and Pattern-recognition, pp. 770–778 (2016)
12. Szegedy, C., et al.: Going deeper with convolutions. In: Proceedings of the IEEE Conference On Computer Vision And Pattern Recognition, pp. 1–9 (2015)
13. Goodfellow, I., et al.: Generative adversarial nets. In: Advances in Neural information Processing Systems, pp. 2672–2680 (2014)
14. Zhu, J.-Y., Park, T., Isola, P., Efros, A.A.: Unpaired image-to-imagetranslation using cycle-consistent adversarial networks. In: Proceedings of the IEEE International Conference on Computer Vision, pp. 2223–2232 (2017)
15. Kim, T., Cha, M., Kim, H., Lee, J. K., Kim, J.: Learning to discovercross-domain relations with generative adversarial networks. In: Proceedings of the 34th International Conference on Machine Learning-Volume 70, pp. 1857–1865. JMLR. org (2017)
16. Isola, P., Zhu, J.-Y., Zhou, T., Efros, A.A.: Image-to-image translation withconditional adversarial networks. In: Proceedings of the IEEE on Computervision and Pattern Recognition, pp. 1125–1134 (2017)
17. Si, C., Wang, W., Wang, L., Tan, T.: Multistage adversarial losses forpose-based human image synthesis. In: Proceedings of the IEEE Conference on Computer Vision and Pattern Recognition, pp. 118–126 (2018)
18. Lassner, C., Pons-Moll, G., Gehler, P.V.: A generative model of peoplein clothing. In: Proceedings of the IEEE International Conference on Computer Vision, pp. 853–862 (2017)
19. Omran, M., Lassner, C., Pons-Moll, G., Gehler, P., Schiele, B.: Neural bodyfitting: Unifying deep learning and model based human pose and shape estimation. In: 2018 International Conference on 3D Vision (3DV), pp. 484–494. IEEE (2018)
20. Loper, M., Mahmood, N., Romero, J., Pons-Moll, G., Black, M.J.: Smpl: A skinned multi-person linear model. ACM Trans. Graph. (TOG), **34**(6), 248 (2015)
21. Tang, W., Li, T., Nian, F., Wang, M.: Mscgan: Multi-scale conditional generative adversarial networks for person image generation. CoRR, abs/1810.08534 (2018)

22. Balakrishnan, G., Zhao, A., Dalca, A. V., Durand, F., Guttag, J.: Synthesizing images of humans in unseen poses. In: Proceedings of the IEEE Conference on Computer Vision and Pattern Recognition, pp. 8340–8348 (2018)
23. Zhu, Z., Huang, T., Shi, B., Yu, M., Wang, B., Bai, X.: Progressive pose attention transfer for person image generation. In: Proceedings of the IEEE Conference on Computer Vision and Pattern Recognition, pages 2347–2356 (2019)
24. Esser, P., Sutter, E., Ommer, B.: A variational u-net for conditional appear-ance and shape generation. In: Proceedings of the IEEE Conference on Computer Vision and Pattern Recognition, pp. 8857–8866 (2018)
25. Kingma, D.P, Welling, M.: Auto-encoding variational bayes (2013) arXiv preprint arXiv:1312.6114
26. Neverova, N., Alp Guler, R., Kokkinos, I.: Dense pose transfer. In: Proceedings of the European Conference on Computer Vision (ECCV), pp. 123–138 (2018)
27. Horiuchi, Y., Iizuka, S., Simo-Serra, E., Ishikawa, H.: Spectral normalizationand relativistic adversarial training for conditional pose generation with self-attention. In: 2019 16th International Conference on Machine Vision Applications (MVA), pp. 1–5. IEEE (2019)
28. Zanfir, M., Popa, A.-I., Zanfir, A., Sminchisescu, C.: Human appear-ance transfer. In: Proceedings of the IEEE Conference on Computer Vision and Pattern Recognition, pp. 5391–5399 (2018)
29. Han, X., Wu, Z., Wu, Z., Yu, R., Davis, L.S.: Viton: An image-basedvirtual try-on network. In: Proceedings of the IEEE Conference on Computer Vision and Pattern Recognition, pp. 7543–7552 (2018)
30. Wang, B., Zheng, H., Liang, X., Chen, Y., Lin, L., Yang, M.: Toward characteristic-preserving image-based virtual try-on network. In: Proceedings of theEuropean Conference on Computer Vision (ECCV), pages 589–604 (2018)
31. Raj, A., Sangkloy, P., Chang, H., Hays, J., Ceylan, D., Lu, J.: SwapNet: image based garment transfer. In: Ferrari, V., Hebert, M., Sminchisescu, C., Weiss, Y. (eds.) ECCV 2018. LNCS, vol. 11216, pp. 679–695. Springer, Cham (2018). https://doi.org/10.1007/978-3-030-01258-8_41
32. Zhao, B., Wu, X., Cheng, Z.-Q., Liu, H., Jie, Z., Feng, J.: Multi-view imagegen-eration from a single-view. In: 2018 ACM Multimedia Conference on Multimedia Conference, pp. 383–391. ACM (2018)
33. Zhou, X., Huang, Q., Sun, X., Xue, X., Wei, Y.: Towards 3d humanpose esti-mation in the wild: a weakly-supervised approach. In: Proceedings of the IEEE International Conference on Computer Vision, pp. 398–407 (2017)
34. Tome, D., Russell, C., Agapito, L.: Lifting from the deep: Convolutional 3dpose estimation from a single image. In: Proceedings of the IEEE Conference on Computer Vision and Pattern Recognition, pp. 2500–2509 (2017)
35. Zhihui, S., Ming, Y., Guohui, Z., Lei, D., Jianda, S.: Cascade Feature Aggregation for Human Pose Estimation (2019). CoRR abs/1902.07837
36. Radford, A., Metz, L., Chintala, S.: Unsupervised representation learning with deep convolutional generative adversarial networks. (2015) arXiv preprint arXiv: 1511.06434
37. Zhang, H., et al.: Stackgan: Text to photo-realistic image synthesis with stacked generative adversarial networks. In: Proceedings of the IEEE International Con-ference on Computer Vision, pp. 5907–5915 (2017)
38. Newell, A., Yang, K., Deng, J.: Stacked hourglass networks for human pose esti-mation. In: Leibe, B., Matas, J., Sebe, N., Welling, M. (eds.) ECCV 2016. LNCS, vol. 9912, pp. 483–499. Springer, Cham (2016). https://doi.org/10.1007/978-3-319-46484-8_29

39. Carreira, J., Agrawal, P., Fragkiadaki, K., Malik, J.: Human pose estimationwith iterative error feedback. In: Proceedings of the IEEE Conference on Computer Vision and Pattern Recognition, pp. 4733–4742 (2016)

40. Quan, T.M., Hildebrand, D.G., Jeong, W.-K.: Fusionnet: A deep fullyresidual convolutional neural network for image segmentation in connectomics. (2016) arXivpreprint arXiv:1612.05360

41. Srisha, R., Khan, A.: Morphological operations for image processing : Understanding and its applications (2013)

42. Mathieu, M., Couprie, C., LeCun, Y.: Deep multi-scale video predictionbeyond mean square error (2015). arXiv preprint arXiv:1511.05440

43. Zheng, L., Shen, L., Tian, L., Wang, S., Wang, J., Tian, Q.: Scalable personreidentification: A benchmark. In: Proceedings of the IEEE International Conference on Computer Vision, pp. 1116–1124 (2015)

44. Liu, Z., Luo, P., Qiu, S., Wang, X., Tang, X.: Deepfashion: Poweringrobust clothes recognition and retrieval with rich annotations. In: Proceedings of the IEEE Conference on Computer Vision and Pattern Recognition, pp. 1096–1104 (2016)

45. Wang, Z., Bovik, A.C., Sheikh, H.R., Simoncelli, E.P., et al.: Image qualityassessment: from error visibility to structural similarity. IEEE Trans. Image Process.$13(4)$:600–612 (2004)

46. Salimans, T., Goodfellow, I., Zaremba, W., Cheung, V., Radford, A., Chen, X.: Improved techniques for training gans. In: Advances in Neural Information Processing Systems, pp. 2234–2242 (2016)

47. Liu, W., et al.: SSD: single shot multibox detector. In: Leibe, B., Matas, J., Sebe, N., Welling, M. (eds.) ECCV 2016. LNCS, vol. 9905, pp. 21–37. Springer, Cham (2016). https://doi.org/10.1007/978-3-319-46448-0_2

48. Everingham, M., Van Gool, L., Williams, C.K., Winn, J., Zisserman, A.: The pascal visual object classes challenge 2007 (voc2007) results (2007)

49. Kingma, D.P. Ba, J.: Adam: A method for stochastic optimization (2014). arXivpreprint arXiv:1412.6980

50. Tieleman, T., Hinton, G.: Lecture 6.5-rmsprop, coursera: Neural networksf or machine learning. University of Toronto, Technical Report (2012)

51. Schaul, T., Zhang, S., LeCun, Y.: No more pesky learning rates. In: International Conference on Machine Learning, pp. 343–351 (2013)

52. Zhang, H., Goodfellow, I., Metaxas, D., Odena, A.: Self-attention generativeadversarial networks (2018). arXiv preprint arXiv:1805.08318

Evaluating Deep Learning Models for the Automatic Inspection of Collective Protective Equipment

Bruno Georgevich Ferreira[✉], Bruno Gabriel Cavalcante Lima,
and Tiago Figueiredo Vieira

Institute of Computing, Federal University of Alagoas, Maceió, Alagoas, Brazil
{bgf,bgcl,tvieira}@ic.ufal.br

Abstract. Deep Learning models are becoming widely used in many applications but they can often be improved through a fine-tuning process capable of enhancing their performances in specific scenarios.

In this paper we tackle the problem of autonomously inspecting the conditions of Collective Protection Equipment (CPE) such as fire extinguishers, warning signs, ground and wall signalization and others.

Work ministry imposes that such CPE are in good conditions to prevent accidents, carrying out periodic mandatory *in loco* auditions. Industry is increasingly applying the potential of Deep Learning (DL) models to automatize such Computer Vision (CV) tasks and a fiber-optic component provider proposed this demand.

Specifically, we assessed the performances of four DL models, namely, MobileNet V2 SSDLite, FPN Resnet-50 SSD, Inception Resnet v2 Faster R-CNN with Atrous Convolution and EfficientNet B0 SSD in the evaluation of CPE conditions. We provide results that highlight each architecture's advantages and drawbacks in the aforementioned scenario.

Indeed, experiments have shown their potential in reducing time and costs of periodic inspections in factories.

Keywords: Deep Learning · Object detection · Visual inspection · Collective Protection Equipment

1 Introduction

Computer Vision tasks are usually related to a domain presenting both high dimension and variability. Nevertheless, Deep Learning (DL) has been successfully applied in many scenarios due to its good capability in dealing with a large amount of high dimension data [6]. But DL models can often be improved through a fine-tuning process capable of enhancing their performances in specific scenarios [1]. This is particularly useful when an appropriate partnership happens between industry and innovation laboratories which stimulates the transfer of knowledge from academia to practical applications in industry.

Supported by organization x.

With respect to the manufacture industry, work ministry imposes that such Collective Protective Equipment (CPE) are in good conditions to prevent accidents, carrying out periodic mandatory *in loco* auditions. To alleviate risk and improve its processes, the industry is increasingly applying the potential of Deep Learning (DL) models to automatize such Computer Vision (CV) tasks and a factory producing fiber-optic components presented such demand. In this paper we tackle the problem of autonomously inspecting the conditions of CPE such as fire extinguishers, warning signs, ground and wall signalization and others. A diagram of the application flow can be found in the Fig. 1.

This paper is an extended version of the paper published in [7] in DeLTA (*Deep Learning Theory and Applications* – 2020). More specifically, we present the following contributions:

1. We collect and annotate on an information base containing attributes specifically concerned with handling the organization's interest. No similar database (containing highlights) has been introduced up until this point to the extent of our knowledge.
2. We survey the performance of four DL models, namely, MobileNet V2 SSDLite, FPN Resnet-50 SSD, Inception Resnet v2 Faster R-CNN with Atrous Convolution and EfficientNet B0 SSD and analyze the compromise among precision and processing time. The system was installed on a tablet that can possibly be attached to a mobile robot. More robust models can also be deployed on machines with higher performance (when compared to tablets) in order to provide a higher confidence level when analyzing surveillance videos accessible through Ethernet.

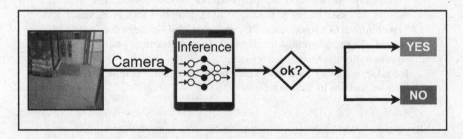

Fig. 1. Overview of the application flow [7].

2   Deep Learning in Industry Works

Artificial Intelligence has been experiencing tremendous advances in the past decade. Tech giants in different industries such as movie streaming, social networks, shared commuting and data networks are taking samples of data as inputs and extracting predictions out of it. This has enabled new products launches and other competitive advantages. Even though companies have already gained awareness of AI's importance, there is still some time to go until they are fully proficient in incorporating DL's functionalities into their processes.

Deep Learning has presenting a pervasive effect in many industries. For instance, authors in [2] have shown that Convolutional Neural Networks have been leading scientific journal publications in the construction industry. Many works have been presented with applications ranging from pavement stress detection and classification, compressive strength and crack prediction of recycled concrete, safety guardrail detection, workforce activity assessment, and others. They also cite several other works using techniques for natural language processing and others using autoencoders.

Hocensky et al., applied well established image processing techniques in three modules aimed at detecting problems in ceramic tiles, particularly those attached to surface, edges and corners [8]. Their proposal achieved a good result in evaluating pottery and was capable of operating in real time.

Battery Management Systems (BMS) are critical in electrical vehicles. It is important to analyze the actual status of the battery, since this module has potential impact in many other control systems of the car. Authors in [19] stated that models of battery estimation were incapable of providing real time results and were too complex. They proposed to apply DL techniques to simplifying the estimator so it could operate in real time. Using highly accurate Neural Network (NN) models capable of approximating the BMS estimation function, the system became capable of operating in real time.

Many devices are being incorporated into local fogs via wireless networks in industry 4.0. Internet of Thing (IoT) is an increasingly growing field and it is becoming much easier to install devices in industry facilities. This improves process monitoring and control since the amount of data becomes larger due to many devices providing information with high frequency. Authors in [11] proposed a way of predicting faulty products by applying DL techniques on a large database. It is important to notice, however, that verification should be done in real time to satisfy industry demands.

Pointing out that autonomous cars should take decision carefully, Rao et al. [15] presented several points regarding the application of Deep Learning in self driving vehicles. Specifically, it is noticeable that, since environmental perception, path planning or even steering wheel control is carried ou by DL-based approaches, they need to undergo strict assessment concerning models' function safety. Such findings are particularly important since autonomous vehicles are transitioning from prototypes to manufacture.

In petroleum industry, the work presented in [20] used YOLOv2 for the automatic detection of oil facilities in order to improve safety in extraction and production. As a baseline, they compared the DL performance with traditionally engineered features, such as, Haar cascades and AdaBoost classification. As expected, YOLOv2 provided a better performance.

Since ship manufacturing is a highly dangerous industry, there are several restrictions, techniques and safety policies to diminish the number of accidents. There is the risk of unpredictable accidents due to the highly dynamic environment added to cases of improper use of personal and collective protection equipment. With that in mind, authors in [4] estimated the inherent risk of the environment. The goal was to propose several safety measures aimed at making the work a safer place.

A detection scheme for defect coffee grain was proposed by [5], using a Generative-Adversarial Network (GAN) which augmented the database and labeled new data. The approach allowed the development of a model with greater generalization capability and also decreased significantly the process of create a database. Consequently, it made the process of training and testing the model much simpler.

3   Methodology

The steps followed in this work are listed below.

1. Constructing a database for each class of labeled images;
2. Developing deep network models that consider architectures that are more suitable for the problem at hand;
3. Checking the models that were trained.

3.1   The Database

Gathering the images was the first step to creating the database. Therefore, in different locations, numerous pictures of fire extinguishers have been taken, such as the factory itself, university laboratories, other buildings, etc. The challenge in locating fire extinguishers that were very rusty or with visible flaws was difficult. This was due to the fact that safety laws which require the replacement of near-to-expire fire extinguishers. Thus, photos, including some samples of defective extinguishers, were also obtained from the internet. Another solution was to take images of numerous rusty items, where only the rust was annotated. Instead of teaching just what rusty extinguishers look like, this approach aimed at teaching the model features of rusty objects.

The annotation process started after the collection of a set of fire extinguisher pictures and emergency lamps. In the extinguisher photos, the hose, the signaling plate, the rusty points, the extinguisher body and the floor signal were annotated. The male socket plug, the female socket plug, the status LED and the lamp body were annotated on the emergency lamp images. These features were chosen by taking into account the demand submitted by the company, which claimed that the most common faults occurring on their facilities are the aforementioned ones.

With respect to the extinguisher, only the hoses in good condition were annotated, given that the damaged hoses (including those positioned incorrectly) were very different and, thus, the model could not detected them. Another explanation for not annotating the circumstances in which the hose was found in a bad position was due to the proximity of the classes in the feature space, resulting in increased classification difficulty. The power plug and the female socket on the wall were annotated with the emergency lamps. We believed that if one of them was found the emergency lamp was not properly wired into the power line and was thus unused.

The database did not comprise too many images per class because it was designed from scratch, which led us to apply some augmentation techniques to minimize overfitting. The first pre-processing approach was to scale all images to pixels of 300×300 and convert them to JPEG, which occupies less space. The following techniques were used for data augmentation;

– vertical and horizontal mirroring;
– 90° rotation;
– bright adjustment;
– resizing
– cropping

During preparation, all augmentation techniques were applied randomly. The bounding boxes were taken into account in the event of resizing and cropping. In order to generate a new sub-image sample, regions containing one or more objects were extracted when multiple objects of interest were present in the same image. This way, with variations of its objects, one picture might create sub-images. For each sub-image, there was at least one object visible.

3.2   Network Architectures

Four neural network architectures with very distinct characteristics were chosen so we could determine the pros and cons of each one. MobileNet V2 SSDLite, suggested by [16], was the first. FPN Resnet-50 SSD, presented by [12], was the second. The third was the Inception Resnet V2 Faster R-CNN with Atrous Convolution, adapted from [17]. The last one was the EfficientNet, proposed by [18], combined with the Single-Shot Multibox Detector, from [14]. Both architectures, for impartial comparison, were trained using the same train and test datasets. All architectures are summarized in Table 1.

Table 1. Summary of Neural Networks Architectures used in this work.

Architecture	Reference
MobileNet V2 SSDLite	[16]
FPN Resnet-50 SSD	[12]
Inception Resnet V2 Faster R-CNN with Atrous Convolution	[17]
EfficientNet and Single-Shot Multibox Detector	[14,18]

MobileNet V2 SSDLite. As the first topology to be tested, MobileNet V2 SSDLite (MV2) was selected because it was designed to run on mobile devices, which makes it faster than other topologies. The most significant contribution provided in MV2 were new layers known as Inverted Residual with Linear Bottleneck, according to [16]. This new layer presents a reduced dimension input, which is first extended to an increased dimension and filtered with separable convolutions that are depth-wise. Next, filtered features have their dimension reduced through linear convolutions. The author also proposes a SSDLite that is a variation of the SSD, proposed by [14], with the convolutional layers being replaced by depth-wise separable convolutions.

Although the most important contribution of MV2 is the new Inverted Residual layers with Linear Bottleneck, they still inherit some very important characteristics from their predecessor: MobileNet V1 ([9]). Depth-wise separable convolutions are the key inherited features, which minimize the required number of mathematical operations in one inference, making the topology faster for training and testing. Depth-wise separable convolutions are the replacement of a factorized version of two different layers for traditional convolution. The first layer is the convolution of the depth-wise, which performs a low-cost convolution applying only one filter per input layer. The second layer is a convolution of 1×1, which is called a point-wise convolution, used to measure new features from the linear input layer combination. This is how separable convolutions are performed in detail. First, per each input layer, it only applies one convolutional filter. It then summarizes the characteristics produced on the previous layer with a linear convolution of 1×1.

FPN Resnet-50 SSD. The FPN Resnet-50 SSD (FPN50) has the following characteristics: it has a pyramid network feature (FPN) as a common feature extractor; it has 50 residual layers; it uses SSD as a multi-box detector.

The FPN is significant because it sums up the model's invariance to scale. As for the Resnet-50, its integration into the chosen model was critical as residual layers allow deeper networks while preventing over-fitting. This is possible since convolutions are used by residual layers only if strictly necessary. If not required, the layer will reproduce the input on the output.

By replacing the Faster-RCNN with the SSD, the FPN Resnet-50, suggested by [12], was altered. The FPN50 was chosen due to its capability of detecting more complex features.

Inception Resnet V2 Faster R-CNN with Atrous Convolution. The Inception Resnet v2 Faster R-CNN with Atrous Convolution (IRV2) was chosen due to the fact that, according to [10], it currently has one of the best results in object detection. Due to its considerable depth it has good learning abilities. On the other hand, it topology has a slower inference time than the previously cited ones.

We use a combination of the Inception Resnet v2 presented in [17] with the Atrous Convolution, proposed in [3]. A model trained with this topology is unable to compete with MV2 and FPN50 with respect to efficiency in time, because the Faster-RCNN is the region's proposed architecture. Faster-RCNN, as shown by [10], offers a higher inference time compared to SSD.

Therefore, the key purpose of using this topology was to test the efficiency of a model well known for its high generalization capability and high standard of estimation and classification of bounding boxes. In order to compare previous models with it, this will introduce a best case scenario.

EfficientNet B0 SSD. Expanding a convnet causes it to increase its accuracy. There are currently three possible expansion methods, by width, depth and resolution. The expansion methods, when applied alone, presented promising results, however they saturated quickly, showing that they allow only a small improvement. In this way, [18] proposed the composite expansion method, which uses the three previous methods combined in order to achieve the maximum possible gain in terms of precision for the model.

The authors claim that the composite expansion method makes sense, intuitively, considering that the insertion of a larger entrance, naturally, requires a deeper and wider model to learn more complex patterns and increase the reception field of the image. network, or receptive field. The authors also proposed a model called EfficentNet, which was built from the application of the NAS method of architecture optimization. The model estimated by the NAS was named EfficientNet B0 (ENB0), and the next versions will be derived from it. The other models will be expanded versions of the B0, using the composite expansion method.

It is assumed that B0 is the smallest and simplest model and B7 the largest and most complex, considering that it has the largest expansion factor, therefore more parameters and therefore more complexity. The authors concluded that the composite expansion method allows the models to focus on more relevant areas with more detailed objects, while the other models allowed for more details or more learning capacity.

So, the combination of EfficientNet B0 with the SSD detector was chosen because they unite the good feature extraction and lightness proposed by EfficientNet with the fastest object detection provided by SSD. This model has a good inference time and it's accuracy is comparable with the IRV2.

3.3   The Training/Test Cycle

For training and testing, it was established that the database augmentation techniques would be used, taking into account the low number of images per class. Variables are selected here, such as the split ratio between training and test sets, as well as measurement indicators for model results. For each model, training time was documented as well.

Augmentation Techniques. The augmentation strategies chosen were: vertical and horizontal mirroring; rotation of $90°$; bright adjustment; resizing; and cropping.

The aim of mirroring and rotation was to generalize the shapes of objects present on the training base. However, the inverse effect can be created by excessive use of this method of manipulation, making the model filters account more for color and texture.

The aim of resizing and cropping was to improve the invariance to scale for the models, which will increase the model's generalization potential and turning the dataset less pronounced for overfitting, since all images found in the dataset will be seen as a different image in the training loop.

This approach was aimed at making the model less sensitive to color with regard to the bright manipulation of the image, accounting more for contour patterns and shapes.

Hardware and Software. Together with Object Detection APIs from TensorFlow V1 and V2, an RTX 2080 Ti video graphics card with 11 GB of GDDR6 RAM was used. The option of a 20XX family GPU from Nvidia to train the models was due to the existence of special cores in it, called TensorCores, which greatly reduced training time.

The object detection Tensorflow API contains all topologies that have been addressed previously, among others. This allows for rapid prototyping, including simple parameter changes. This API also provides well-known data-sets with pre-trained models, enabling techniques such as transfer learning. For small custom databases, it is extremely useful.

For the three topologies on the MSCOCO dataset, developed by [13], pre-trained models were used in the adopted training phase. We used 75 and 25 percent of the training and research ratio database, respectively. The experiments were carried out with a Snapdragon 845 processor on an Android smartphone.

Metrics. The following metrics have been established for the evaluation of training and testing time models: training set loss; test set loss; Average Recall (AR) and Mean Average Precision (mAP) in the test set; AR Across Scales and mAP Across Scales in the test set; frame rate per second (FPS) on mobile devices and on RTX 2080 Ti.

Loss analysis, both on training and test collection, focused on determining if the models were generalizing well or whether any over-fitting characteristics were presented. There are three key configurations for the AR metric calculation: (1) one detection per image (AR@1); (2) ten detections per image (AR@10); and (3) 100 detections per image (AR@100). AR studies that use more detections appear to be higher.

Related to the AR, there are three major versions of mAP calculations: (1) the mean Average Precision (AP) over the Intersection over Union (IoU) limits, with values between 50% and 95% and steps of 5%; (2) the mean AP with IoU limits set at 50% (mAP@0.5IoU); (3) the mean AP with IoU limits set at 75% (mAP@0.75IoU).

For the Across Scales metrics, three sets of objects are calculated: (1) small objects with an area less than 32^2 pixels; (2) medium objects with an area between 32^2 and 96^2 pixels; and (3) large objects with dimensions greater than 96^2 pixels. Thus, the AR Across Scales is calculated for 100 detections of small objects (AR@100 small), medium objects (AR@100 medium) and large objects (AR@100 large).

mAP Across Scales, on the other hand, is measured using the first mAP settings for small objects (mAP small), medium objects (mAP medium) and large objects (mAP large). The FPS indicator analysis, consisting of the number of images per second that the model is capable of analyzing, is often used to analyze speed and performance.

4   Results and Discussion

As the preparation and research exercises were being carried out, findings were obtained and analyzed, determining whether any configurations needed to be changed. Thus, in this section, obtained results from each methodology section will be presented.

4.1   Image Dataset

The database created has 137 images of emergency lamps, 147 of rusty objects and 256 of fire extinguishers. The emergency lamps images were not used to train the models. This was attributed to the fact that some of the emergency class lamp artifacts did not show very distinctly, which was the same issue with the led status class. The power socket from the emergency lamps could not be identified by trained models either, explained by the lack of appropriate photos for proper generalization. We therefore decided to train the models for fire extinguishers only for the classes strictly related. Some examples from the custom dataset are shown in Fig. 2.

4.2   Trained Models Metrics

Considering the metrics discussed in Sect. 3, output for each of the trained models will be evaluated in this section.

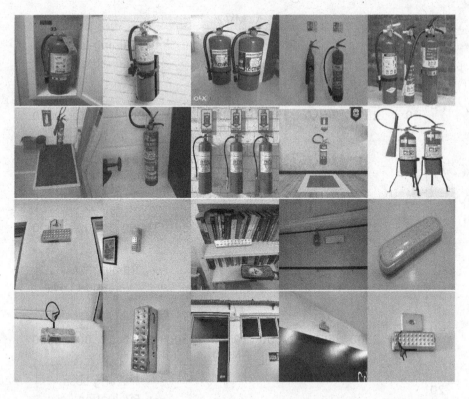

Fig. 2. Fire extinguishers database (top couple rows), vertical (red signs showing where the extinguishers are) and horizontal (yellow stripes) signs and emergency lights (bottom two rows) [7]. (Color figure online)

Losses. Losses for each model can be seen in Fig. 3 for both training and testing sets. The findings show that losses are substantially lower for FPN50, EBN0 and IRV2 than for MV2.

Average Recall. The AR results using all the configurations discussed are shown here: AR@1, AR@10 and AR@100. You can see these in the Fig. 4. It shows that MV2 had comparable IRV2 performance, although it took more time to achieve its benefit. With regard to FPN50, considering the initial steps of training, it showed better results but had poorer performance relative to the other two models at the end of training. Throughout the training, ENB0 maintained better than other models, no matter what configurations were used.

Mean Average Precision. Better results were obtained from the mAP@0.5IoU configuration with respect to the mAP metric shown in Fig. 5. For all configurations, considering the last stages of training, the MV2 model was able to approach the IRV2 from a broader perspective. Once again, FPN50 presented the best results at the start of the training, but did not manage to hold its lead until the end. ENB0 maintained itself as the best model through all the preparation, no matter what configuration.

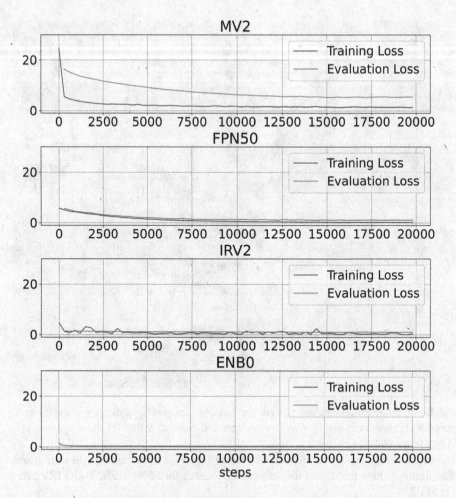

Fig. 3. Training and evaluation losses.

Average Recall Across Scales. The AR Across Scales metric results are shown in the Fig. 6. All models presented similar AR values, for all object sizes. IRV2 showed the best performance overall, while MV2 presented the slowest but most stable growth. All models presented their best AR values when analyzing large-sized objects.

For small-sized objects, AR results were very similar at the end, with MV2 showing the smallest AR. During the training, ENB0 presented some high peaks, finishing with the best values. IRV2 and FPN50 finished both as the second best models, with almost exactly the same AR values. MV2 performed worse than the others at the end.

Analyzing medium-sized objects, it is possible to notice that ENB0 reached convergence first, while MV2 presented itself as the slowest. FPN50 showed more oscillations during the training process. At the end, all models reached very close AR values. IRV2 and ENB0 presented slightly better AR values during final training steps, with FPN50 showing the worst performance at the end.

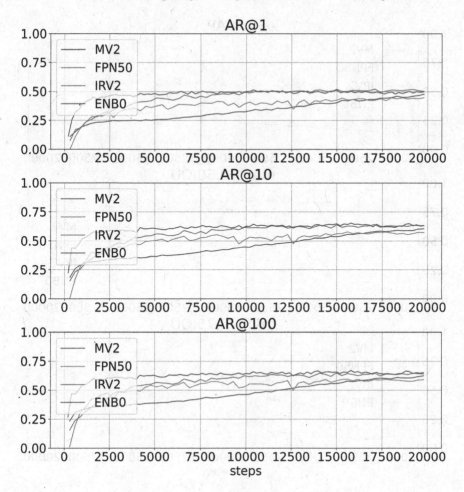

Fig. 4. Average Recall.

With respect to large-sized objects, models performed better that other sizes. ENB0 showed a better growth at first steps, but was rapidly outperformed by IRV2. While presenting slow growth during all the training, MV2 finished as the best model, presenting results very close to IRV2 model. FPN50 and ENB0 performed worse, also presenting similar final AR values.

Mean Average Precision Across Scales. As shown in Fig. 7, the findings showed that MV2 did not produce good results for small artifacts, performing its best for large objects. On the other hand, MV2 presented the most stable learning across all artifact sizes. IRV2 model presented good results across all object sizes. ENB0 was the best performing model for medium-sized objects, but performed poorly of small objects.

For small objects, FPN50 and IRV2 showed to be the best performing models, both with similar mAP values. MV2 showed better stabilization during all the training,

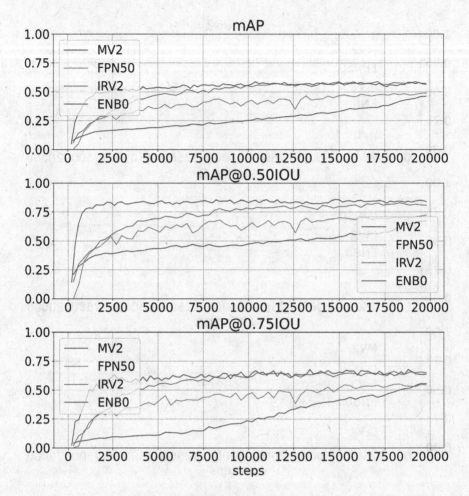

Fig. 5. Mean Average Precision calculated.

staying as the second best model at the end. ENB0 presented more oscillations at the end, finishing the training with the smallest mAP values.

For medium objects, FPN50 and MV2 performed worse. The ENB0 model showed the best mean precision overall. IRV2 was the second best model in this category at the end, performing simillar to FPN50 at the first stages of training, but performing significantly better ant the end. FPN50 and MV2 were the worst for this size of objects.

Analyzing large objects, IRV2 and ENB0 presented themselves, again, as the best models, although results were very close at the final steps of training. Analyzing the convergence speed at the start of the training, it is possible to notice that ENB0 is the fastest while MV2 is the slowest. On the other hand, MV2 presented the larger growth of mAP at the end, potentially reaching the other models mAP values with more time steps.

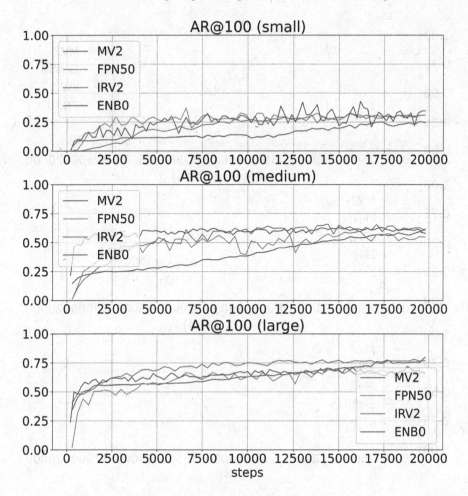

Fig. 6. Average Recall Across Scales.

Table 2. FPS rate in RTX 2080 Ti and mobile device.

Models	RTX 2080 Ti	Mobile Device
MV2	42	5
ENB0	16	1
FPN50	12	–
IRV2	3	-

Frames per Second Ratio. As seen in Table 2, it can be shown that MV2 produces much better results with respect to Frames per Second (FPS) when compared to the other three models. The only one that is able to run continuously on a mobile device is the MV2. The ENB0 is capable of running on mobile devices with 1 FPS, allowing for intermittent inferences only. The third fastest is the FPN50. Finally, IRV2, operating on

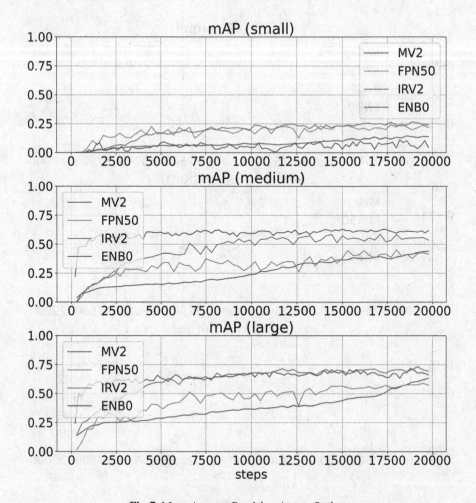

Fig. 7. Mean Average Precision Across Scales.

a 3 FPS average, is not capable of achieving good performance, even when running on a RTX 2080.

Discussions About Models Performance. The MV2 model proved to be very flexible, presenting good performance with both medium and large-sized objects. On the other hand, the ENB0 model showed to be light and fast, while achieving results comparable to those of the IRV2 model, including in cluttered environments. Such characteristics make ENB0 a complete model.

The IRV2 model was slower than the others, but presented the best results overall. MV2 and FPN50 did not perform as well in cases where the input image contained objects in a more cluttered environment, despite achieving good results in all other cases.

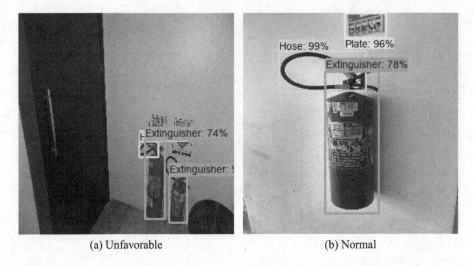

(a) Unfavorable   (b) Normal

Fig. 8. Performances of FPN50's topology when submitted to unfavorable and normal scenarios [7].

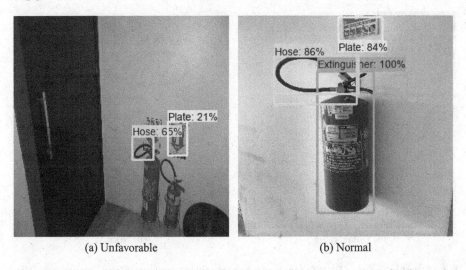

(a) Unfavorable   (b) Normal

Fig. 9. Performances of MV2's topology when submitted to unfavorable and normal scenarios [7].

Analyzing the images shown in the Figs. 8a, 9a, 11a and 10a, it can be seen that IRV2 and ENB0 were able to capture more complex image patterns. They performed successfully even when the pictures were taken from unaligned angles. Examples of these not well-behaved cases can be seen in Figs. 8a, 9a, 11a and 10a.

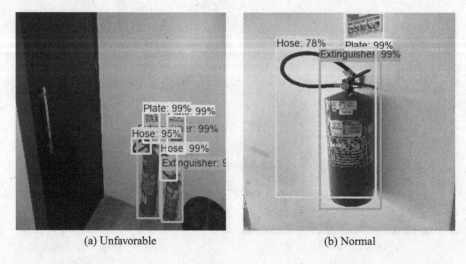

(a) Unfavorable   (b) Normal

Fig. 10. Performances of IRV2's topology when submitted to unfavorable and normal scenarios [7].

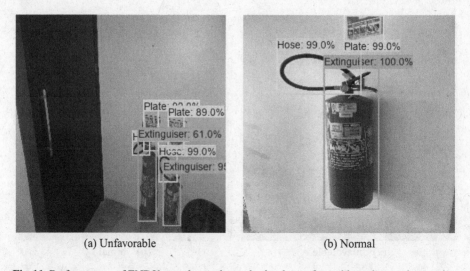

(a) Unfavorable   (b) Normal

Fig. 11. Performances of ENB0's topology when submitted to unfavorable and normal scenarios.

5   Conclusion and Future Works

Four models were identified which were capable of detecting faults in fire extinguishers. Used methodology can be extended to other artifacts from the industry environment, potentially being applied to different forms of auditing. We also managed to collect a particular database, through using diversified sources which include, but are not limited to, taking new photos of existing extinguishers. Data augmentation was also used. Such database was used for the training process of our models.

Results showed that the MV2 model enables auditing to be carried out in real time when executed on mobile devices. This model showed to be the fastest among tested ones. Executing it on computers leads to better performances, if needed.

The ENB0 showed to be quite promising, being capable of performing on mobile devices while maintaining itself with the best mAP. The FPN50 was capable of detecting small, medium and large defects in fire extinguishers, besides enabling real-time execution, but was not able to be executed on a mobile device. When relating to speed, it was characterized as an in-between model.

IRV2 showed the ability to detect more complex patterns, including a better detection of extinguishers' defects. The total numbers of false-positives and false-negatives for this model were significantly smaller. On the other hand, IRV2 required more powerful processing capacity to be carried out. Using pre-training upon larger datasets, along with transfer-learning, reduced the total training time, making it easier for the model to converge. For deep learning applications, this sort of approach is acceptable.

As future works, we plan to increase both the size of our dataset and the number of classes within it. An example, we aim to collect emergency lamp's photographs, so that their audit can be done along with extinguishers and other CPE items.

We also plan to address other industry concerns, such as verification of extinguishers' load and pressure, and whether its labeling panel is present and readable. More up-to-date topologies will also be tested upon our study case, in search of better performance - mobile devices included. For instance, we aim to test CenterNet object detector, which has been presenting good results recently.

Acknowledgements. The authors would like to thank the Edge laboratory http://edgebr.org/ at the Federal University of Alagoas https://ufal.br/ which funded this research through an agreement regulated by Brazil's Information Technology Law.

References

1. Aggarwal, C.C.: Neural Networks and Deep Learning, 1 edn. Springer International Publishing (2018). https://doi.org/10.1007/978-3-319-94463-0, https://www.springer.com/gp/book/9783319944623
2. Akinosho, T.D., et al.: Deep learning in the construction industry: A review of present status and future innovations. J. Building Eng. **32**, 101827 (2020). https://doi.org/10.1016/j.jobe.2020.101827
3. Chen, L.C., Papandreou, G., Kokkinos, I., Murphy, K., Yuille, A.L.: Deeplab: Semantic image segmentation with deep convolutional nets, atrous convolution, and fully connected crfs. IEEE Trans. Pattern Anal. Mach. Intell. **40**(4), 834–848 (2017)
4. Choi, Y., Park, J.H., Jang, B.: A risk estimation approach based on deep learning in shipbuilding industry. In: 2019 International Conference on Information and Communication Technology Convergence (ICTC), pp. 1438–1441. IEEE (2019)
5. Chou, Y.C., et al.: Deep-learning-based defective bean inspection with gan-structured automated labeled data augmentation in coffee industry. Appl. Sci. **9**(19), 4166 (2019)
6. Dargan, S., Kumar, M., Ayyagari, M.R., Kumar, G.: A survey of deep learning and its applications: a new paradigm to machine learning. Arch. Comput. Methods Eng. **27**(4), 1071–1092 (2019). https://doi.org/10.1007/s11831-019-09344-w

7. Georgevich Ferreira, B., Lima., B.G.C., Vieira., T.F.: Visual inspection of collective protection equipment conditions with mobile deep learning models. In: Proceedings of the 1st International Conference on Deep Learning Theory and Applications - Volume 1: DeLTA, pp. 76–83. INSTICC, SciTePress (2020). https://doi.org/10.5220/0009834600760083

8. Hocenski, Ž., Matić, T., Vidović, I.: Technology transfer of computer vision defect detection to ceramic tiles industry. In: 2016 International Conference on Smart Systems and Technologies (SST), pp. 301–305. IEEE (2016)

9. Howard, A.G., et al.: Mobilenets: Efficient convolutional neural networks for mobile vision applications. arXiv preprint arXiv:1704.04861 (2017)

10. Huang, J., et al.: Speed/accuracy trade-offs for modern convolutional object detectors. In: Proceedings of the IEEE Conference on Computer Vision and Pattern Recognition, pp. 7310–7311 (2017)

11. Li, L., Ota, K., Dong, M.: Deep learning for smart industry: Efficient manufacture inspection system with fog computing. IEEE Trans. Industr. Inf. **14**(10), 4665–4673 (2018)

12. Lin, T.Y., Dollár, P., Girshick, R., He, K., Hariharan, B., Belongie, S.: Feature pyramid networks for object detection. In: Proceedings of the IEEE Conference On Computer Vision and Pattern Recognition, pp. 2117–2125 (2017)

13. Lin, T.-Y., et al.: Microsoft COCO: common objects in context. In: Fleet, D., Pajdla, T., Schiele, B., Tuytelaars, T. (eds.) ECCV 2014. LNCS, vol. 8693, pp. 740–755. Springer, Cham (2014). https://doi.org/10.1007/978-3-319-10602-1_48

14. Liu, W., et al.: SSD: single shot multibox detector. In: Leibe, B., Matas, J., Sebe, N., Welling, M. (eds.) ECCV 2016. LNCS, vol. 9905, pp. 21–37. Springer, Cham (2016). https://doi.org/10.1007/978-3-319-46448-0_2

15. Rao, Q., Frtunikj, J.: Deep learning for self-driving cars: chances and challenges. In: Proceedings of the 1st International Workshop on Software Engineering for AI in Autonomous Systems, pp. 35–38 (2018)

16. Sandler, M., Howard, A., Zhu, M., Zhmoginov, A., Chen, L.C.: Mobilenetv 2: Inverted residuals and linear bottlenecks. In: Proceedings of the Ieee Conference On Computer Vision and Pattern Recognition, pp. 4510–4520 (2018)

17. Szegedy, C., Ioffe, S., Vanhoucke, V., Alemi, A.A.: Inception-v4, inception-resnet and the impact of residual connections on learning. In: Thirty-First AAAI Conference on Artificial Intelligence (2017)

18. Tan, M., Le, Q.V.: Efficientnet: Rethinking model scaling for convolutional neural networks. arXiv preprint arXiv:1905.11946 (2019)

19. Veeraraghavan, A., Adithya, V., Bhave, A., Akella, S.: Battery aging estimation with deep learning. In: 2017 IEEE Transportation Electrification Conference (ITEC-India), pp. 1–4. IEEE (2017)

20. Zhang, N., et al.: Automatic recognition of oil industry facilities based on deep learning. In: IGARSS 2018–2018 IEEE International Geoscience and Remote Sensing Symposium, pp. 2519–2522. IEEE (2018)

Intercategorical Label Interpolation for Emotional Face Generation with Conditional Generative Adversarial Networks

Silvan Mertes[✉], Dominik Schiller, Florian Lingenfelser, Thomas Kiderle,
Valentin Kroner, Lama Diab, and Elisabeth André

University of Augsburg, Universitätsstraße 1, 86159 Augsburg, Germany
{silvan.mertes,dominik.schiller,florian.lingenfelser,
thomas.kiderle,valentin.kroner,lama.diab,
elisabeth.andre}@informatik.uni-augsburg.de

Abstract. Generative adversarial networks offer the possibility to generate
deceptively real images that are almost indistinguishable from actual photographs. Such systems however rely on the presence of large datasets to realisti-
cally replicate the corresponding domain. This is especially a problem if not only
random new images are to be generated, but specific (continuous) features are to
be co-modeled. A particularly important use case in *Human-Computer Interaction* (HCI) research is the generation of emotional images of human faces, which
can be used for various use cases, such as the automatic generation of avatars.
The problem hereby lies in the availability of training data. Most suitable datasets
for this task rely on categorical emotion models and therefore feature only discrete annotation labels. This greatly hinders the learning and modeling of smooth
transitions between displayed affective states. To overcome this challenge, we
explore the potential of label interpolation to enhance networks trained on categorical datasets with the ability to generate images conditioned on continuous
features.

Keywords: Generative adversarial networks · Face generation · Conditional
GAN · Emotion generation · Label interpolation

1 Introduction

With recent advances in the field of *Generative Adversarial Learning*, a variety of new
algorithms have emerged to address artificial image data generation. The state of the art
Generative Adversarial Networks (GANs) are characterized by high image quality of
generated results compared to other generative approaches such as *Variational Autoencoders*. However, early GAN architectures lack the ability to generate new data in a
controllable way. The original GAN framework has been modified and extended in a
variety of ways in order to enable such a controlled generation of new images. These
modified architectures have demonstrated the ability to address a broad range of image
generation tasks. Especially in the field of *Human-Computer Interaction* (HCI), these
systems are a promising tool. One particularly relevant task is the generation of avatar
images, which are images of human faces that can be controlled with respect to various

A. Fred et al. (Eds.): DeLTA 2020/DeLTA 2021, CCIS 1854, pp. 67–87, 2023.
https://doi.org/10.1007/978-3-031-37320-6_4

human-interpretable features. In the context of emotional face generation, this enables the generation of avatar images conditioned on a particular emotion.

Most datasets suitable for training such face generation GANs refer to categorical emotion models, i.e., they contain emotion labels that were annotated in a discrete way, e.g., the emotions refer to emotional states like happy or sad. However, for many real-world use cases, such as emotional virtual agents, corresponding face images need to be generated in a more detailed manner to improve the credibility and anthropomorphism of human-like avatars. This is especially of interest during the design stage of such virtual agents, as the consistency of different modalities of virtual agents is of great importance [10, 27] and fine-grained degrees of expressivity can enhance the perception of certain affective states of the agent, influencing (among others) the perception of the agent's personality [15].

Moreover, images that only show single emotions are not realistic in situations where smooth transitions between different emotional states are required. Other use cases include automatically creating textures for virtual crowd generation or augmenting data for emotion recognition tasks. Especially in the latter case, there is a huge need for artificially created data, since continuous emotion recognition relies on non-categorical training data, and available datasets labeled in terms of dimensional features are scarce. In all these cases, the use of dimensional emotion models would be more sufficient to meet the requirements posed.

In this work, we explore the applicability of label interpolation for *Conditional GANs* (cGANs) that were trained on categorical datasets. By doing so, we study the possibility to bypass the need for continuously labeled datasets. Since categorical labels are essentially binned version of continuous labels, it makes sense that the samples belonging to a specific categorical label are covering a large spectrum of expressiveness. We believe that this information can be learned by a generative model and being exploited to create emotional images on a continuous scale. To explore the feasibility of our hypothesis, we first train cGANs on two datasets widely used for benchmarking various deep learning tasks, namely CIFAR-10 [17] and Fashion-MNIST [47]. Those datasets contain discrete class labels that we use for conditioning the GAN. We then examine the effects of interpolating between those discrete class labels, by observing how a pre-trained classifier behaves when looking at continuously interpolated results. From the insights gained from these more generic datasets, we tackle the concrete use-case of emotional face generation as already described in [28]. By doing so, we enable the cGAN to generate faces showing emotional expressions that can be controlled in a continuous, dimensional way. The goal of this paper is therefore to answer the question of whether label interpolation can be a tool to overcome the drawbacks of categorical datasets for emotional face generation.

Extending our already published work [28], this paper explores not only on the applicability of label interpolation to the scenario of emotional face generation, but additionally reports on preceding experiments (see Sect. 4), gaining more in-depth insights into the feasibility of label interpolation.

2 Background

When Generative Adversarial Networks (GANs) were first introduced by Goodfellow et al. [11] they sparked a plethora of research innovations in the field of artificial data generation. GANs are based on the idea of two neural networks competing against each other in a min-max game. While the *Generator* network aims to generate new data that has never been observed before and imitates the original training domain as closely as possible, the *Discriminator* tries to discern original samples from the target domain and fake sample created by the generator. As a result, the generator learns to produce artificial data samples resembling the original domain from a random noise vector. Over the last years, this basic concept has been refined and advanced considerably, aiming to improve the training and output performance of GANs in various ways [1, 2, 12]. Along with setting the new state-of-the-art of quality for artificial image generation, GANs have opened up new possibilities for facial image generation and modification.

[32], for instance, modified the original GAN architecture by exchanging the fully-connected layer architecture with convolutional networks in the generator and discriminator, allowing, among others, to generate high quality human face images. Additionally, they investigated how their so-called *Deep Convolutional GAN* (DCGAN) implicitly maps the latent space to facial features (e.g. face pose). However, since the training procedure of DCGAN is unsupervised, it comes with the inherent drawback that the facial feature types to be learned can not be directly controlled. [14] presented *Progressive Growing GANs*, an even more sophisticated approach to adversarial image generation, that they applied to the task of human face generation.

The aforementioned approaches all underlay the drawback that their outputs solely depend on the random noise input vector, without the possibility to control it in a human readable way. This problem was addressed by [30], who use *Conditional GANs* (cGAN) to encode additional label information in the input vector, enabling the network to consider certain pre-defined features in the output. This property of cGANs was exploited by [9,46] to generate face image data with respect to specific features (e.g. *glasses*, *gender*, *age*, *mouth openness*). Similarly, [48] made use of the cGAN conditioning mechanisms in order to augment emotional face image datasets. One problem of this approach is the usage of either using discretely labeled features, restricting the output to discrete categories, or to already use continuously labeled data during training which is rarely available in a plethora of scenarios.

A related task that GANs are frequently applied to is the task of *Style Conversion*, which in terms of facial expressions is also known as *Face Editing*. It intends to modify existing image data instead of generating entirely new data [5, 13, 20, 21, 35]. Using GANs, [6] managed to develop a framework that allows to continuously adapt the emotional expressions of images. Although their approach is not explicitly based on continuously annotated data, the diversity of the intensity of emotions must be represented in the training set. Their system proved its capability of generating random new faces expressing a particular emotion. However, they didn't investigate the generation capabilities of their system according to common known dimensional emotion models like Russel's Valence-Arousal model [36]. The focus was rather to show that their face editing system is able to modify the intensity of discrete, categorical emotions.

In general, interpolating through the label space of a cGAN is a quite under-explored mechanism. Direct manipulation of the latent input space of GANs has been made possible by various automated approaches like *Latent Vector Evolution* [41] as well as interactive ones [40]. Also, in generative approaches apart from adversarial learning, exploring interpretable and non-interpretable latent spaces are a widespread tool, for example in the context of Human-Robot Interaction [34] or Speech Synthesis [33]. However, manipulating the discrete label space of cGANs in a continuous way has not yet found its way into practice. To the best of the authors' knowledge, there is no system that is trained on discrete emotion labels and outputs new face images that can be controlled in a continuous way.

3 Technical Framework for Interpolating Categorical Labels

In order to explore the applicability of label interpolation in cGANs, an appropriate framework had to be defined, which is presented in the following sections.

3.1 Network Architecture

The networks utilized in our experimental settings are largely founded on a *Deep Convolutional GAN* (DCGAN) by [32]. A detailed description of the original DCGAN architecture can be found in the respective publication. In summary, DCGANs are a modification to the original GAN framework by [11], where convolutional and convolutional-transpose layers were included in order to model the training domain with higher image quality. The architectures that we used were modified to fit the corresponding datasets. Additionally, to enable targeted image generation (which is not part of the original DCGAN), the architectures were extended with the principles of a *cGAN*.

Unlike conventional GANs, cGANs incorporate a conditioning mechanism consisting of an additional class input vector. This vector is used to control specific features of the output images by telling the generator network about the presence of certain features during training. Thus, this feature information must be given as labels while training the cGAN. Thus, the input for a cGAN consists of a random noise component z (as in the original GAN framework) and a conditioning vector v. After the training process, the generator has learned to transform the random noise input into images that resemble the training domain, taking into account the conditioning information given by v in order to drive the outputs to show the desired features.

In our implementation, the conditioning information is given to the network as one-hot encoded label vector, where each element represents a certain feature. Thus, the one-hot label vector v has the following form [28]:

$$v = (v_1, v_2, ..., v_n) = \{0, 1\}^n \tag{1}$$

where n is the number of controlled features. The datasets that we used in our experiments are primarily designed for classification tasks. This implies that we consider a feature a class of the dataset. As, in the scope of this work, only datasets for single-class classification were considered, the following restriction holds true [28]:

$$\sum_{i=1}^{n} v_i = 1 \tag{2}$$

3.2   Interpolation

After training, the definition of the condition part of the cGAN's input vector is changed to allow for a continuous interpolation between the originally discrete classes. Generally, this can simply be done by reformulating the conditining vector v so that is not forced to a binary structure [28]:

$$v = (v_1, v_2, ..., v_n) = [0, 1]^n \tag{3}$$

During our experiments, we found that keeping the restriction formulated in Eq. 2 leads to better quality of interpolated results instead of picking the single elements of the vector arbitrarily in the interval $[0, 1]$. In other words, interpolation is done by subtracting some portion e from the input representative of one class and adding it to another class. Our hypothesis is that due to the differentiable function that is approximated by the cGAN model during the training process, those non-binary conditioning vectors lead to image outputs which are perceived as lying somewhere *between* the original, discrete classes. For our target context, the generation of face images with continuous emotional states, this would refer to images of faces that do not show the extreme, discrete emotions that are modeled in a categorical emotion system, but to more fine-grained emotional states as they are conventionally modeled by a dimensional emotion model as will be further elaborated on in Sect. 5.1.

4   Feasibility Studies

To evaluate the feasibility of our approach, we decided to first apply it to two generic datasets, before finally addressing the problem of emotional human face generation.

4.1   Datasets

The Fashion-MNIST dataset [47] encompasses a set of product pictures taken from the Zalando website, where each image belongs to one of 10 classes. Each of these contains 7,000 pictures. The images that we used are 8-bit grayscale versions with a resolution of 28×28 pixels. All in all, this results in a dataset of 70,000 fashion product pictures, whereas 60,000 are attributed to the training dataset and 10,000 to the test set. Examples for each class are depicted by Fig. 1.

The CIFAR-10 and the CIFAR-100 datasets both are derived from the *80 million tiny images dataset* [17]. In contrast to the 100 classes of CIFAR-100, CIFAR-10 only contains a subset of 10 classes, whereas each class has 6,000 colored images of size 32×32. This results in a dataset of 60,000 images in total, where 50,000 belong to the training and 10,000 to the test set. The classes are mutually exclusive, even for narrow classes like trucks and cars. Figure 2 depicts example images for the corresponding 10 classes.

We decided to use the Fashion-MNIST dataset because it has originally been designed for measuring the performance of machine learning approaches. The pictures

T-Shirt

Trouser

Pullover

Dress

Coat

Sandal

Shirt

Sneaker

Bag

Ankle Boot

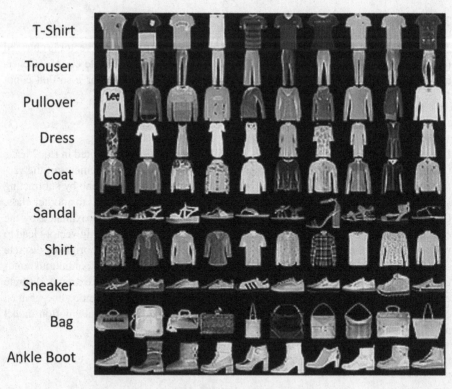

Fig. 1. Fashion-MNIST categories and examples. [47].

are grayscaled and comparably small, making the dataset suitable for preliminary feasibility experiments. To further test the viability of our approach, we aimed to increase the challenge gradually. Thus, we additionally chose to use the CIFAR-10 dataset. Although it also contains small pictures, the challenge is raised by the colorization and the slightly higher resolution.

4.2 Methodology

In order to evaluate if the interpolation algorithm creates smooth transitions between two arbitrary classes, we decided to perform a fine-grained analysis on the continuously generated outputs by the use of our approach. To this end, we used pre-trained classifiers that are able to accurately distinguish between the different discrete classes contained in the respective datasets. As the focus of this work is to gain insights into the question whether interpolating between discrete label information can be a promising tool for future applications, the discrete decision of such classification models are not a good metric for our purposes. Instead, we want to explore if the interpolation mechanism is able to model the full bandwidth of transitional states that can occur *between* different classes. Thus, for evaluating if the interpolation mechanism works correctly, we assessed the confidence of the classification models that the interpolated result belongs to certain classes. Ideally, during interpolation, this confidence should continuously shift towards the class that is interpolated to.

Airplane
Automobile
Bird
Cat
Deer
Dog
Frog
Horse
Ship
Truck

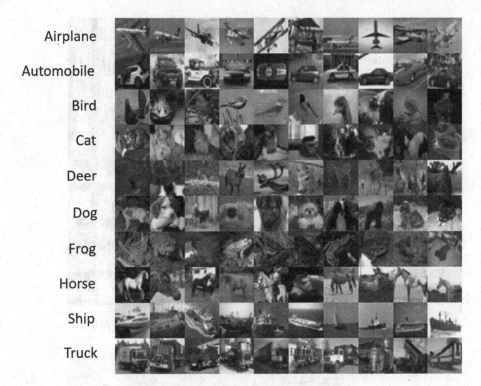

Fig. 2. CIFAR-10 categories and examples [17].

4.3 Training

For both the datasets, we adapted the DCGAN architecture to fit the dataset. Slight changes to the architecture had to be made in order to produce reasonable outputs. Further, we enhanced both models with the conditioning mechanism as described in Sect. 3.1.

Fashion-MNIST. For this dataset, we trained the cGAN model for 20,000 random batches of size 32 on all of the 50,000 images of the *train* partition of the dataset using Adam optimizer with a learning rate of 0.0002 and β_1 of 0.5. Example outputs of the trained model can be seen in Fig. 3, whereas example outputs of different interpolation steps are shown in Fig. 5.

CIFAR-10. For this dataset, we trained the cGAN model for 30,000 random batches of size 32 on all of the 50,000 images of the *train* partition of the dataset, again using Adam optimizer with a learning rate of 0.0002 and β_1 of 0.5. Example outputs of the trained model can be seen in Fig. 4, whereas example outputs of different interpolation steps are shown in Fig. 6. In both the images, it can be clearly seen that the chosen cGAN architecture apparently was not able to resemble the traing domain sufficiently enough. Results are blurry, and objects can only partially be recognized as the intended objects. However, we chose to continue with the validation of the interpolation as we

T-Shirt
Trouser
Pullover
Dress
Coat
Sandal
Shirt
Sneaker
Bag
Ankle Boot

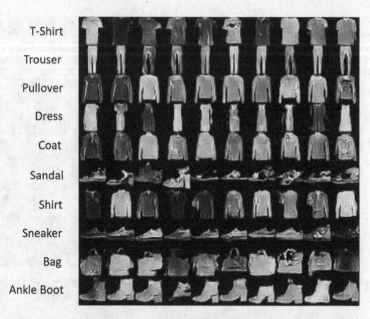

Fig. 3. Exemplary outputs of the cGAN model trained on Fashion-MNIST.

Airplane
Automobile
Bird
Cat
Deer
Dog
Frog
Horse
Ship
Truck

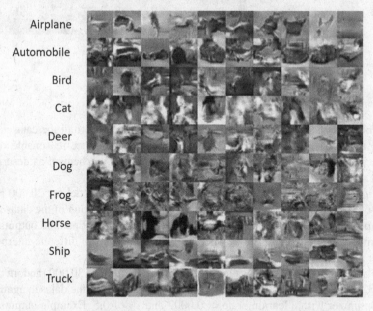

Fig. 4. Exemplary outputs of the cGAN model trained on CIFAR-10.

were also interested in how label interpolation behaves when dealing with models that do not represent the respective training domain very well.

T-Shirt to coat

Trouser to shirt

Pullover to dress

Dress to trouser

Coat to sandal

Sandal to bag

Shirt to pullover

Sneaker to ankle boot

Bag to sneaker

Ankle Boot to T-Shirt

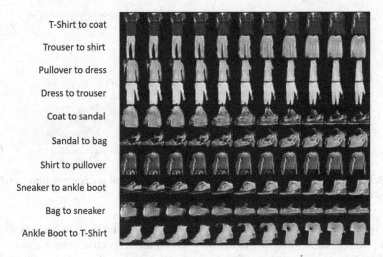

Fig. 5. Exemplary outputs of the interpolation steps of the cGAN model trained on Fashion-MNIST.

Airplane to deer

Automobile to dog

Bird to horse

Cat to bird

Deer to truck

Dog to cat

Frog to automobile

Horse to ship

Ship to frog

Truck to airplane

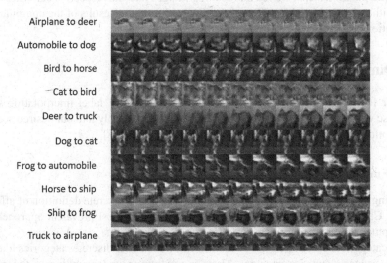

Fig. 6. Exemplary outputs of the interpolation steps of the cGAN model trained on CIFAR-10.

4.4 Computational Evaluation

In order to test the capability to interpolate between different classes, we used classifiers that we trained on the task of object classification. To this end, we used the EfficientNet-B0 architecture [42], as these models turned out to achieve very high accuracy on both datasets (*Fashion-MNIST:* 0.9089, *CIFAR-10:* 0.9931). We used a softmax layer on top of the models, which produces an output vector $r \in I\!R^{+\ n}$ with $\sum_{i=1}^{n} r_i = 1$ where n is the number of classes. By interpreting this class probability vector r as confidence distribution over all the classes, we can assess the interpolation capabilities of the cGAN models by observing the change of r. To this end, 1,000 image sets were

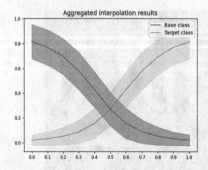
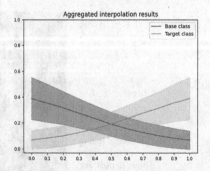

Fig. 7. Results of the computational evaluation with Fashion-MNIST.

Fig. 8. Results of the computational evaluation with CIFAR-10.

randomly generated for each class combination i, j in CIFAR-10 as well as Fashion-MNIST. Each of these images was conditioned on the respective source class i. Then, we performed interpolation steps for every source image as described in Sec. 3.2 with $\alpha = 0.1$, resulting in 10 interpolation steps until the target class was reached. For each interpolation steps, we fed all resulting images into the respective classifier model (i.e., either the Fashion-MNIST or the CIFAR-10 model). Results of the computational evaluation are plotted in Fig. 7 and Fig. 8.

5   Dimensional Face Generation

As our feasibility studies revealed, that the mechanism of label interpolation shows promise when being used with more generic datasets, we apply it to our desired scenario of emotional face generation, as we already described in [28].

5.1   Emotion Models

Enabling algorithms to handle human emotion requires a discrete definition of affective states. Categorical and dimensional models are the two most prevalent approaches to conceptualize human emotions.

A categorical emotion model subsumes emotions under discrete categories like happiness, sadness, surprise or anger. There is a common understanding of these emotional labels, as terms describing the emotion classes are taken from common language. It is also for this reason, that labels are the more common form of annotation found with datasets depicting emotional states. However, this (categorical) approach may be restricting, as many blended feelings and emotions cannot adequately be described by the chosen categories. Selection of some particular expressions can not be expected to cover a broad range of emotional states, especially not differing degrees of intensity. An arguably more precise way of describing emotions is to attach the experienced stimuli to continuous scales within dimensional models. [25] suggests to characterize emotions along three axes, which he defines as pleasure, arousal and dominance. [18] proposes the simplified axes of arousal and valence as measurements, resulting in the more commonly used dimensional emotion model. The valence scale describes the pleasantness of a given emotion. A positive valence value indicates an enjoyable emotion such as joy or pleasure. Negative values are associated with unpleasant emotions like sadness and fear. This designation is complemented by the arousal scale which measures the agitation level of an emotion (Fig. 9). This representations is less intuitive but allows continuous blending between affective states.

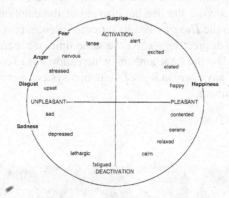

Fig. 9. Russel's 2-dimensional valence arousal circumplex [36].

Categorical as well as dimensional models are simplified, synthetic descriptions of human emotions and are not able to cover all of the included aspects. However, with our interpolation approach we aim to cover all the whole emotional range defined within the space of the dimensional valence-arousal model and enable a seamless transition between displayed emotions. As data collections featuring dimensional annotation for facial expressions are more sparse than the ones containing categorical labels (Sect. 5.2), being able to use emotional labels in the training process is very beneficial. Goal of the following study is to use a cGAN that was conditioned on categorical emotions during training, and interpolate between those emotions in order to be able to create new images. Those newly generated face images show emotional states that are located in the continuous dimensional space of the valence/arousal model without having to correlate directly with discrete emotion categories.

To formally represent the valence and arousal of a face image I, we use a tuple $VA(I) = (v, a)$, where v refers to valence and a to arousal. Correlating with Russel's theory explained above, an image x with $VA(x) = (0, 0)$ is representing the center of the emotion space and thus show a neutral emotion. Emotions that are referred to in categorical emotion systems (e.g., *Happy*, *Sad*) are represented by valence/arousal states that show quite extreme values. When it comes to the interpolation of those dimensional emotional states, i.e., to create images with certain degrees of arousal or valence, we interpolate between the *neutral* emotion and the *extreme* emotional states. By the term *extreme emotion*, we refer to all categorical emotional states used except the *neutral* state, as this represents the center of the dimensional emotion model.

In our experiments, we stuck to performing interpolations between *Neutral* and a particular other emotion to preserve comparability between emotions. It should be noted that the approach could easily be extended to interpolate between two or even more categorical emotions. However, since we use only one categorical emotion and *Neutral* at a time, the following restriction must be added:

$$\exists i_{\in [2,6]} : v_1 + v_i = 1 \tag{4}$$

where v_1 represents the condition for *Neutral*.

To create an image that should show a specific degree of valence v or arousal a, where $0 \leq a, v \leq 1$, we use the one-hot element of the emotion that maximizes the specific value, for example *Happy* when it comes to valence, or *Angry* for arousal, and then decrease it to the desired degree. At the same time, we increase the one-hot element related to *Neutral* by the same amount, which allows us to create images showing valence/arousal values anywhere in Russel's emotion system, as opposed to the *extreme* values given during training.

5.2 Dataset

Fig. 10. Exemplary data from FACES showing neutral, sad, disgust, fear, anger and happiness from left to right varying the age group [7].

As previously mentioned, datasets labeled in terms of dimensional emotional models are scarce. Although there are a few datasets with continuous labeled information (e.g. *AffectNet* by [31] or AFEW-VA by [16]), they use to be gathered in the wild, resulting in miscellaneous data.

Data diversity usually is beneficial for deep learning tasks, however, in our specific use case of face generation with the focus on modeling certain emotional states in human faces, consistency in all non-relevant characteristics (i.e., characteristics not related to facial expressivity) is an advantage.

Thus, although a variety of categorically labeled datasets are available [3, 19, 22, 24, 38, 43], we decided to use the FACES dataset [7] for our experiments, since it meets our requirements particularly well. In this dataset all images are labeled in a discrete manner, and recorded with an identical uniformly coloured background and an identical grey shirt. This is exemplified in Fig. 10. To overcome the disadvantages of continuously labeled, but inconsistently recorded emotional face datasets, we explore the use of label interpolation with categorically labeled datasets.

Overall the FACES dataset consists of 2052 emotional facial expression images, distributed over 171 men and women. The 58 participants are assigned to the group young, 56 to middle-aged and 57 to the old group, each showing 2 styles of the emotions *Neutral, Fear, Anger, Sadness, Disgust* and *Happiness*. For training we only needed to resize the pictures to a target resolution of 256×256 pixels.

5.3 Methodology

As our feasibility study revealed, the interpolation approach has potential for creating transitions between different discrete states. However, it could be seen that the quality of the generated images, especially when dealing with the CIFAR-10 dataset, left room for improvement. To use the approach of label interpolation in a real world scenario like avatar generation or similar, such a poor image quality would be unacceptable. Thus, besides optimizing the cGAN model for our face generation use case even more, our evaluation process here is two-folded. First, we evaluate whether the cGAN is, before applying any interpolation, able to create images that are perceived correctly by human judgers. By doing so we can assess if the cGAN model that we trained is capable of generating images with sufficient enough quality to express emotional states. Secondly, we conducted a computational evaluation analogously to the feasibility study.

Neutral   Sadness   Disgust   Fear   Anger   Happiness

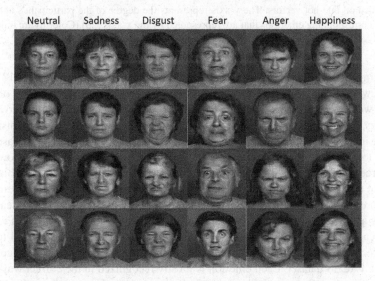

Fig. 11. Example outputs of the trained cGAN model [28].

5.4 Training

The model was trained for 10,000 epochs on all 2052 images of the FACES dataset using *Adam* optimizer with a learning rate of 0.0001. Example outputs of the trained model, conditioned on one-hot vectors of all 6 used emotions, are shown in Fig. 11.

5.5 User Evaluation

In our user study, we evaluated the cGAN's ability to produce images of discrete emotions generated with the respective one-hot vector encoding. In total, 20 probands of ages ranging from 22 to 31 years (M = 25.8, SD = 2,46, 40% male, 60% female) participated in the study.

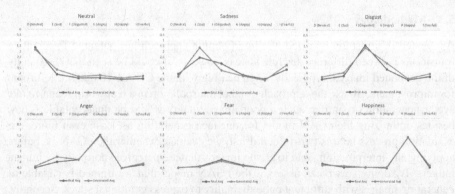

Fig. 12. Results of the user study. Blue graphs show the perceived emotion of real images from the FACES dataset, while orange graphs show the perceived emotion of outputs of the cGAN conditioned on one-hot vectors. The y-axis represents the degree of the participant's agreement with the corresponding emotions that are represented by the x-axis [28] (Color figure online).

During the survey, 36 images were shown to each of the participants. 18 of the images where original images taken from the FACES dataset, whereas the other 18 images were generated by the trained cGAN. All images were split evenly between all emotions, both for the original as well as for the generated images. To keep consistency with the images generated by the cGAN, the images taken from the FACES dataset were resized to 256×256 pixels. For each image, the participants were asked how much they agreed to the image showing a certain emotion. To mitigate confirmation bias, they were not told which emotion the image should show, but asked to provide their rating for each emotion. The ratings were collected by the use of a 5-point Likert scale (1 = strongly agree, 5 = strongly disagree). Results of the user study are shown in Fig. 12.

As can be seen, the images that were generated by the cGAN were rated to show the respective targeted emotion in a similar convincing way as the original images taken from the FACES dataset. Each emotion is mostly recognized in the correct way by the study participants. One emotion, namely *Sadness*, even stands out as the artificially generated images were recognized even better than the original images, which were mistaken for *Disgust* more frequently. Considering these results, the trained cGAN model proves to be an appropriate basis for interpolation experiments.

5.6   Computational Evaluation

Analogously to the computational evaluation in our feasibility studies, we verified if label interpolation can be used to enhance the cGAN network with the ability to generate images with continuous degrees of valence and arousal with the help of an auxiliary classifier. Again, 1,000 noise vectors per class were initially fed into the cGAN, where here, the classes were the five emotions *Sadness, Disgust, Fear, Anger* and *Happiness*. The conditioning vector was initially chosen to represent the neutral emotion. For each of the 5,000 noise vectors, 10 interpolation steps with step size $e = 0.1$ towards the respective extreme emotion were conducted. Thus, the last interpolation step results in a one-hot vector representing the respective extreme emotion. For evaluting the resulting valence/arousal values, we again used a pre-trained auxiliary classifier.

Figure 13 shows exemplary outputs of interpolation steps between *neutral* and the five used emotions.

Table 1. AffectNet performance comparison [28].

	AffectNet Baseline		Evaluation Model	
	Valence	Arousal	Valence	Arousal
RMSE	0.37	0.41	0.40	0.37
CORR	0.66	0.54	0.60	0.52
SAGR	0.74	0.65	0.73	0.75
CCC	0.60	0.34	0.57	0.44

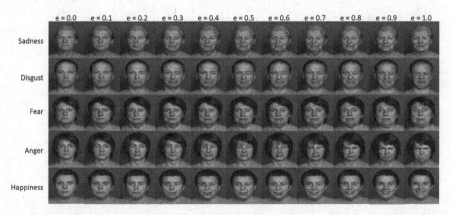

Fig. 13. Example outputs of the interpolation mechanism. Each row shows a set of interpolation steps, where in each step, the emotion portion e was increased by 0.1, whereas the neutral portion was decreased by the same amount [28].

The auxiliary classifier model was based on the MobileNetV2 architecture [37]. The model was trained on the AffectNet dataset for 100 epochs with *Adam* optimizer and a learning rate of 0.001, leading to a similar performance as the AffectNet baseline models, as can be seen in Table 1. We assessed the valence/arousal values for every interpolated output image of the cGAN and averaged them over the 1,000 samples per emotion, analogously to the feasibility studies described in Sect. 4.4. The results can be taken from Fig. 14.

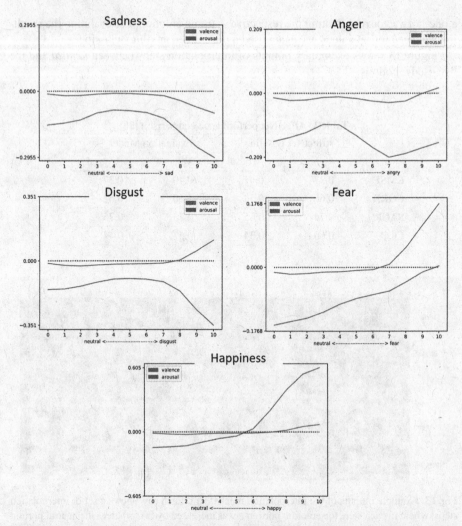

Fig. 14. Computational Evaluation of our interpolation approach. Red graphs show valence, while green graphs show arousal. The x-axis represents the interpolation steps. Each interpolation step was performed by increasing the corresponding emotion vector element by 0.1, while decreasing the neutral vector element by 0.1 [28] (Color figure online).

6   Discussion

In our initial study we evaluated if our proposed approach can be used to seamlessly interpolate images between generic classes. To this end we relied on two widely used and publicly available datasets CIFAR-10 and Fashion-MNIST (see 4.1) to train our cGAN interpolation model. Figure 5 and Fig. 6 are showing examples of the calculated interpolations between various classes on the Fashion-MNIST dataset and the CIFAR-10 dataset respectively. We can clearly see that the trained network was not able to capture the complexity of the input domain optimally. While the images from the Fashion-MNIST domain are showing slightly blurred contours, the generated images from the CIFAR-10 domain can only be partially recognized as the intended objects. However, when looking at the individual morphing steps between the classes, we can observe that the model is able to generate transitions that are generally smooth and continuous - two necessary prerequisites to apply the approach to interpolate between emotional expressions in human faces, in order to create meaningful results.

To further validate this observation we also employed a trained classifier for each dataset to predict the various interpolation steps between classes. Assuming a well calibrated classifier, we expected the distribution of the predicted class probabilities to continuously shift between the two interpolated classes along with the degree of interpolation.

Figure 7 and Fig. 8 are showing the results for those classifiers as described in Sect. 4.4. In those plots, we averaged the class probabilities for both the base classes and the target classes for every interpolation step of all the assessed output images. It can be observed that, generally speaking, the interpolation mechanism led the network to generate transitions that are indeed perceived as *lying between the classes* by the classifier. Notably, the images produces by the cGAN that was trained on CIFAR-10 were generally classified with quite low confidence. This implies that the assumption that the cGAN model was not able to accurately resemble the dataset holds true. However, for both models, the confidence smoothly transitions between the two intended classes, indicating that label interpolation is a promising tool for further experiments.

We argue that those findings further substantiate the ability of our trained cGAN to generate continuous interpolations between images and therefore also the feasibility to further investigate if the approach is able to generate meaningful interpolations between different categorical emotions.

Upon visual inspection of the Fashion-MNIST and CIFAR-10 datasets, we found that the quality of the artificially generated images from the random noise vectors lagged significantly behind the original samples from the respective areas.

We therefore firstly conducted a user study to assess the capabilities of our employed cGAN model to produce realistic images of people expressing clearly identifiable emotions (see Sec. 5.5. The results of this study, as depicted in Fig. 12, show that participants generally recognize the expressed emotions in the artificially generated images similarly well as in the original images from the FACES dataset. The only exception being the emotion *Sadness*, which was even better identifiable from the artificially generated images than the original data, where participants confused the emotion more often with *Disgust*. Those results are leading us to the conclusion that our employed cGAN model is suitable to further explore interpolation between emotional classes.

The results of the computational evaluation are depicted in Fig. 14 for each emotional class respectively. We can see that the interpolation mechanism is able to condition the cGAN to produce face images with various valence/arousal values. Upon further inspection we can observe that those values are mostly located in the value range between the start and end point of the interpolation, which indicates the general trend of the system to transition smoothly between emotional states. However, the plots also show that the interpolation function is not in all cases strictly monotonic. For example, in the *Sadness* and *Disgust* cases, the valence value initially rises slightly before dropping towards the interpolation endpoint. Similarly for *Anger*, both valence and arousal values are first moving up and down before arriving at their initial starting point. This is a strong deviation to the position of anger in the circumplex model of emotions, where we would expect both valence and arousal to be notably higher when compare with the neutral position. Furthermore, we can see that the detected valence value is in all cases a bit below zero for the neural emotion. Since all emotions have been correctly recognized by human raters we attribute this behaviour to shortcomings in the valence arousal regressor. Taking those human quality ratings and the predominantly correct trend lines of the interpolation into account we argue that our approach can indeed be used to generate face images of continuous emotional states. The fact, that the values are not evolving in a linear way, i.e., the plots appear rather as curves than as straight lines, does not take away much from the results, since the single interpolation step intervals can easily be modified to achieve a more even interpolation. E.g., instead of using the same step interval for every single interpolation step, higher intervals can be used in ranges where the target features are changing slower.

7   Conclusion and Outlook

In this paper, we examined the possibilities of continuous interpolation through a discrete label space of Conditional Generative Adversarial Networks. Therefore, we first conducted some feasibility studies to assess the general applicability of interpolating between discrete classes to a trained cGAN. We found that indeed the technique can be used to generate smooth transitions between classes, even in cases where the cGAN did not learn to model the training domain to a satisfactory level. Subsequently, we applied the label interpolation mechanism to the scenario of continuous emotional face generation. After ensuring that a cGAN trained on a dataset of categorical emotional face images learned to model that categorical emotional states by conducting a user study, we assessed the applicability of label interpolation in order to generate face images that show continuous emotional states. By using an auxiliary classifier for evaluating the cGAN outputs, we found that the algorithm was able to cover most of the valence/arousal ranges that are needed to cover the full dimensional emotion space. Although the performance of the approach shows to be highly dependent on the emotions that are used for interpolation, it shows great potential for application in various use cases such as automatic generation of virtual avatars or crowd generation. In future work, it seems promising to apply label interpolation to GAN models with higher complexity in order to improve the quality of the generated results. Also, it is conceivable to use the proposed system for the task of data augmentation. Previous work has shown that GANs in general have the ability to enhance datasets in order to improve various deep learning tasks, such as semantic segmentation of images [4,29,39,44] or various image- and audio-based classification problems [8,23,26,45]. The ability to abstract from discrete classes to continuous features opens up a variety of machine learning problems where label interpolation could improve performance through data augmentation, which we plan to study in further research.

Acknowledgements. This work has been funded by the European Union Horizon 2020 research and innovation programme, grant agreement 856879.

References

1. Arjovsky, M., Bottou, L.: Towards principled methods for training generative adversarial networks. arXiv preprint arXiv:1701.04862 (2017)
2. Arjovsky, M., Chintala, S., Bottou, L.: Wasserstein generative adversarial networks. In: International Conference on Machine Learning, pp. 214–223. PMLR (2017)
3. Beaupré, M., Cheung, N., Hess, U.: The montreal set of facial displays of emotion, Montreal, Quebec, Canada (2000)
4. Choi, J., Kim, T., Kim, C.: Self-ensembling with gan-based data augmentation for domain adaptation in semantic segmentation. In: Proceedings of the IEEE/CVF International Conference on Computer Vision, pp. 6830–6840 (2019)
5. Choi, Y., Choi, M., Kim, M., Ha, J.W., Kim, S., Choo, J.: Stargan: unified generative adversarial networks for multi-domain image-to-image translation. In: Proceedings of the IEEE Conference on Computer Vision and Pattern Recognition, pp. 8789–8797 (2018)
6. Ding, H., Sricharan, K., Chellappa, R.: Exprgan: facial expression editing with controllable expression intensity. In: Proceedings of the AAAI Conference on Artificial Intelligence, vol. 32 (2018)
7. Ebner, N.C., Riediger, M., Lindenberger, U.: Faces-a database of facial expressions in young, middle-aged, and older women and men: Development and validation. Behav. Res. Methods **42**(1), 351–362 (2010)
8. Frid-Adar, M., Klang, E., Amitai, M., Goldberger, J., Greenspan, H.: Synthetic data augmentation using gan for improved liver lesion classification. In: 2018 IEEE 15th International Symposium on Biomedical Imaging (ISBI 2018), pp. 289–293. IEEE (2018)
9. Gauthier, J.: Conditional generative adversarial nets for convolutional face generation. In: Class Project for Stanford CS231N: Convolutional Neural Networks for Visual Recognition, Winter semester, vol. 2014, no. 5, p. 2 (2014)
10. Gong, L., Nass, C.: When a talking-face computer agent is half-human and half-humanoid: Human identity and consistency preference. Human Commun. Res. **33**(2), 163–193 (2007)
11. Goodfellow, I.J., et al.: Generative adversarial networks. arXiv preprint arXiv:1406.2661 (2014)
12. Gulrajani, I., Ahmed, F., Arjovsky, M., Dumoulin, V., Courville, A.: Improved training of wasserstein gans. arXiv preprint arXiv:1704.00028 (2017)
13. He, Z., Zuo, W., Kan, M., Shan, S., Chen, X.: Attgan: facial attribute editing by only changing what you want. IEEE Trans. Image Process. **28**(11), 5464–5478 (2019)
14. Karras, T., Aila, T., Laine, S., Lehtinen, J.: Progressive growing of gans for improved quality, stability, and variation. arXiv preprint arXiv:1710.10196 (2017)
15. Kiderle, T., Ritschel, H., Janowski, K., Mertes, S., Lingenfelser, F., André, E.: Socially-aware personality adaptation. In: 2021 9th International Conference on Affective Computing and Intelligent Interaction Workshops and Demos (ACIIW). IEEE (2021)
16. Kossaifi, J., Tzimiropoulos, G., Todorovic, S., Pantic, M.: Afew-va database for valence and arousal estimation in-the-wild. Image Vision Comput. **65**, 23–36 (2017)
17. Krizhevsky, A., Hinton, G.: Learning multiple layers of features from tiny images (technical report). University of Toronto (2009)
18. Lang, P.J., Bradley, M.M., Cuthbert, B.N.: Motivated attention: affect, activation, and action. In: Lang, P.J., Simons, R.F., Balaban, M.T. (eds.) Attention and Orienting: Sensory and Motivational Processes, pp. 97–135. Psychology Press (1997)

19. Lang, P.J., Bradley, M.M., Cuthbert, B.N., et al.: International affective picture system (IAPS): technical manual and affective ratings. NIMH Center Study Emot. Attent. **1**, 39–58 (1997)

20. Lin, J., Xia, Y., Qin, T., Chen, Z., Liu, T.Y.: Conditional image-to-image translation. In: Proceedings of the IEEE Conference on Computer Vision and Pattern Recognition, pp. 5524–5532 (2018)

21. Liu, M.Y., Breuel, T., Kautz, J.: Unsupervised image-to-image translation networks. arXiv preprint arXiv:1703.00848 (2017)

22. Lucey, P., Cohn, J.F., Kanade, T., Saragih, J., Ambadar, Z., Matthews, I.: The extended cohn-kanade dataset (ck+): A complete dataset for action unit and emotion-specified expression. In: 2010 IEEE Computer Society Conference on Computer Vision and Pattern Recognition-Workshops, pp. 94–101. IEEE (2010)

23. Mariani, G., Scheidegger, F., Istrate, R., Bekas, C., Malossi, C.: Bagan: data augmentation with balancing gan. arXiv preprint arXiv:1803.09655 (2018)

24. Matsumoto, D.R.: Japanese and Caucasian facial expressions of emotion (JACFEE). University of California (1988)

25. Mehrabian, A.: Framework for a comprehensive description and measurement of emotional states. Genet. Social Gener. Psychol. Monographs **121**(3), 339–361 (1995)

26. Mertes, S., Baird, A., Schiller, D., Schuller, B.W., André, E.: An evolutionary-based generative approach for audio data augmentation. In: 2020 IEEE 22nd International Workshop on Multimedia Signal Processing (MMSP), pp. 1–6. IEEE (2020)

27. Mertes, S., Kiderle, T., Schlagowski, R., Lingenfelser, F., André, E.: On the potential of modular voice conversion for virtual agents. In: 2021 9th International Conference on Affective Computing and Intelligent Interaction Workshops and Demos (ACIIW). IEEE (2021)

28. Mertes, S., Lingenfelser, F., Kiderle, T., Dietz, M., Diab, L., André, E.: Continuous emotions: exploring label interpolation in conditional generative adversarial networks for face generation. In: Fred, A.L.N., Sansone, C., Madani, K. (eds.) Proceedings of the 2nd International Conference on Deep Learning Theory and Applications, DeLTA 2021, Online Streaming, 7–9 July 2021, pp. 132–139. SCITEPRESS (2021). https://doi.org/10.5220/0010549401320139

29. Mertes, S., Margraf, A., Kommer, C., Geinitz, S., André, E.: Data augmentation for semantic segmentation in the context of carbon fiber defect detection using adversarial learning. In: Fred, A.L.N., Madani, K. (eds.) Proceedings of the 1st International Conference on Deep Learning Theory and Applications, DeLTA 2020, Lieusaint, Paris, France, 8–10 July 2020, pp. 59–67. ScitePress (2020). https://doi.org/10.5220/0009823500590067

30. Mirza, M., Osindero, S.: Conditional generative adversarial nets. arXiv preprint arXiv:1411.1784 (2014)

31. Mollahosseini, A., Hasani, B., Mahoor, M.H.: Affectnet: a database for facial expression, valence, and arousal computing in the wild. IEEE Trans. Affect. Comput. **10**(1), 18–31 (2017)

32. Radford, A., Metz, L., Chintala, S.: Unsupervised representation learning with deep convolutional generative adversarial networks. arXiv preprint arXiv:1511.06434 (2015)

33. van Rijn, P., et al..: Exploring emotional prototypes in a high dimensional TTS latent space. In: Proceedings of Interspeech 2021, pp. 3870–3874 (2021). https://doi.org/10.21437/Interspeech.2021-1538

34. Ritschel, H., Aslan, I., Mertes, S., Seiderer, A., André, E.: Personalized synthesis of intentional and emotional non-verbal sounds for social robots. In: 2019 8th International Conference on Affective Computing and Intelligent Interaction (ACII), pp. 1–7. IEEE (2019)

35. Royer, A., et al.: XGAN: unsupervised image-to-image translation for many-to-many mappings. In: Singh, R., Vatsa, M., Patel, V.M., Ratha, N. (eds.) Domain Adaptation for

Visual Understanding, pp. 33–49. Springer, Cham (2020). https://doi.org/10.1007/978-3-030-30671-7_3

36. Russell, J.A., Barrett, L.F.: Core affect, prototypical emotional episodes, and other things called emotion: dissecting the elephant. J. Pers. Social Psychol. **76**(5), 805 (1999)

37. Sandler, M., Howard, A., Zhu, M., Zhmoginov, A., Chen, L.: Mobilenetv 2: inverted residuals and linear bottlenecks. In: 2018 IEEE/CVF Conference on Computer Vision and Pattern Recognition, pp. 4510–4520 (2018). https://doi.org/10.1109/CVPR.2018.00474

38. Van der Schalk, J., Hawk, S., Fischer, A.: Validation of the Amsterdam dynamic facial expression set (adfes). Poster for the International Society for Research on Emotions (ISRE), Leuven, Belgium (2009)

39. Scherer, S., Schön, R., Ludwig, K., Lienhart, R.: Unsupervised domain extension for night-time semantic segmentation in urban scenes (2021)

40. Schlagowski, R., Mertes, S., André, E.: Taming the chaos: exploring graphical input vector manipulation user interfaces for gans in a musical context. In: Audio Mostly 2021, pp. 216–223 (2021)

41. Schrum, J., Gutierrez, J., Volz, V., Liu, J., Lucas, S., Risi, S.: Interactive evolution and exploration within latent level-design space of generative adversarial networks. In: Proceedings of the 2020 Genetic and Evolutionary Computation Conference, pp. 148–156 (2020)

42. Tan, M., Le, Q.: Efficientnet: rethinking model scaling for convolutional neural networks. In: International Conference on Machine Learning, pp. 6105–6114. PMLR (2019)

43. Tottenham, N.: Macbrain face stimulus set. John D. and Catherine T. MacArthur Foundation Research Network on Early Experience and Brain Development (1998)

44. Uricar, M., et al.: Let's get dirty: gan based data augmentation for camera lens soiling detection in autonomous driving. arXiv preprint arXiv:1912.02249 (2019)

45. Waheed, A., Goyal, M., Gupta, D., Khanna, A., Al-Turjman, F., Pinheiro, P.R.: Covidgan: data augmentation using auxiliary classifier GAN for improved Covid-19 detection. IEEE Access **8**, 91916–91923 (2020)

46. Wang, Y., Dantcheva, A., Bremond, F.: From attributes to faces: a conditional generative network for face generation. In: 2018 International Conference of the Biometrics Special Interest Group (BIOSIG), pp. 1–5. IEEE (2018)

47. Xiao, H., Rasul, K., Vollgraf, R.: Fashion-mnist: a novel image dataset for benchmarking machine learning algorithms. CoRR abs/1708.07747 (2017). http://arxiv.org/abs/1708.07747

48. Yi, W., Sun, Y., He, S.: Data augmentation using conditional GANs for facial emotion recognition. In: 2018 Progress in Electromagnetics Research Symposium (PIERS-Toyama), pp. 710–714. IEEE (2018)

Forecasting the UN Sustainable Development Goals

Yassir Alharbi[1,3](✉) ⓘ, Daniel Arribas-Bel[2] ⓘ, and Frans Coenen[1] ⓘ

[1] Department of Computer Science, The University of Liverpool, Liverpool L69 3BX, UK
{yassir.alharbi,coenen}@liverpool.ac.uk
[2] Department of Geography and Planning, The University of Liverpool, Liverpool L69 3BX,
UK
d.arribas-bel@liverpool.ac.uk
[3] Almahd College, Taibah University Al-Madinah Al-Munawarah, Medina, Saudi Arabia

Abstract. This paper presents a review and in-depth analysis of the Sustainable Development Goal Track, Trace, and Forecast (SDG-TTF) framework for
UN Sustainable Development Goal (SDG) attainment forecasting. Unlike earlier
SDG attainment forecasting frameworks, the SDG-TTF framework considers the
possibility for causal relationships between SDG indicators, both within a given
geographic entity (intra-entity relationships) and between the current entity and
its neighbouring geographic entities (inter-entity relationships). The difficulty lies
in identifying such causal linkages. Six different mechanisms are considered. The
discovered causal relationships are then used to generate multivariate time series
prediction models within a bottom-up SDG prediction taxonomy. The overall
framework was assessed using three different geographical regions. The results
demonstrated that the Extended SDG-TTF framework was capable of producing
better predictions than competing models that do not account for the possibility
of intra and inter-causal linkages.

Keywords: Time series causality · Missing values · Hierarchical classification ·
Time series forecasting · Sustainable Development Goals

1 Introduction

On 8 September 2000, at the end of the United Nations (UN) Millennium Summit,
world leaders adopted eight Millennium Development Goals (MDGs) to be achieved
before 2015. The goals are listed in Table 1. Most of the specified goals were achieved
by most countries [28], and the MDG initiative was declared to be a success.

The success of MDG initiative paved the way for another set of goals. In September
2015, the UN introduced the Sustainable Development Goals (SDGs), listed in Table 2,
to be achieved by 2030 [27]. However, this time the goals covered a broader range of
domains. The vision was that achieving these goals would provide for a world free from
hunger and poverty and ensure the sustainability of natural resources and the protection of the environment. The philosophical underpinning for the SDGs initiative, and
the MDG initiative, was the idea that the world is a connected place and that all UN
members should therefore work together to ensure the attainment of these goals for all
member states [8, 22].

ⓒ The Author(s), under exclusive license to Springer Nature Switzerland AG 2023
A. Fred et al. (Eds.): DeLTA 2020/DeLTA 2021, CCIS 1854, pp. 88–110, 2023.
https://doi.org/10.1007/978-3-031-37320-6_5

Table 1. The eight 2000 Millennium Development Goals (MDGs) [3].

1	To eradicate extreme poverty and hunger
2	To achieve universal primary education
3	To promote gender equality and empower women;
4	To reduce child mortality
5	To improve maternal health
6	To combat HIV/AIDS, malaria, and other diseases
7	To ensure environmental sustainability
8	To develop a global partnership for development

Table 2. The seventeen 2005 Sustainable Development Goals (SDGs).

1	No Poverty
2	Zero Hunger
3	Good Health and Well-being
4	Quality Education
5	Gender Equality
6	Clean Water and Sanitation
7	Affordable and Clean Energy
8	Decent Work and Economic Growth
9	Industry, Innovation and Infrastructure
10	Reduced Inequality
11	Sustainable Cities and Communities
12	Responsible Consumption and Production
13	Climate Action
14	Life Below Water
15	Life on Land
16	Peace and Justice Strong Institutions
17	Partnerships to Achieve the Goals

Given the foregoing, predicting whether geographic regions will meet their SDGs or not is of significant interest. Of note is work directed at using machine learning to predict SDG attainment [2,4,21,24]. Most of this existing work assumes that the various SDG targets and indicators are independent of one another. However, this is clearly not the case. The SDGs can be categorised into 4 different levels, as shown in Fig. 1:

Biosphere. environmental-related goals.
Society. Goals related to empowering society, such as by: (i) eradicating poverty, (ii) promoting health and equality and (iii) promoting sustainable cities.
Economy. Goals related to economic growth using responsible consumption while reducing workforce inequality.
Partnership. Goal 17, which exists to promote a global effort to ensure that the SDGs are attained with respect to all countries.

By considering the SDGs in terms of the above categorisation, it can be seen that SDGs are related and connected. For example, without sustained clean water and sanitation (SDG 6), food preparation will be harder; thus, it will affect the attainability of SDG 2

(Zero Hunger); and therefore, SDG 8 (Decent work and economic growth) will not be fulfilled. Given this interconnected nature of the SDGs one can see that there exists a causal relation between individual SDGs.

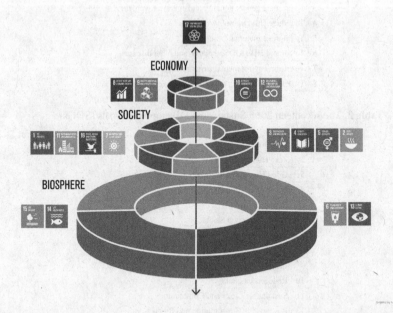

Fig. 1. Interconnectedness in SDG courtesy of Azote and the Stockholm Resilience Centre, Stockholm University [22].

In [3] the SDG Multivariate Track, Trace and Forecast (SDG-TTF) framework was presented that took into consideration both intra-entity relationships and inter-geographic region causalities between SDGs. The proposed SDG-TTF model incorporates the hierarchical framework from [2], and the ACA causality relationship mechanism from [4] for intra- and inter-entity relationship discovery.

This paper presents a much more comprehensive evaluation of the SDG-TTF framework presented in [3] by considering: (i) three different mechanisms for addressing the missing data problem, (ii) three different mechanisms to address the scaling issue that exists within the SDG data, (iii) six different mechanisms for discovering causal relationships and (iv) using a much more substantial portion of the available SDG data to evaluate the framework than originally used in [3], data covering North Africa, East Asia and Northern Europe. The paper also presents a more detailed analysis of the required pre-processing and the adopted mathematical representation.

The rest of this paper is organised as follows. In the following section, Sect. 2, a brief literature review of relevant work underpinning the work presented in this paper is given. The SDG application domain and the SDG time series data set is described in Sect. 3 together with the required pre-processing of the SDG data. The SDG-TTF framework is described in Sect. 4 and its evaluation in Sect. 5. The framework's operation is given in Sect. 6. The paper concludes with a summary of the main findings, a number of proposed directions for future research, in Sect. 7. All the data provided in this paper can be found in the project Github repository[1].

2 Literature Review

The proposed SDG-TTF approach addresses two fundamental challenges: (i) short time series forecasting and (ii) time series causal inference. Previous work in these two areas is therefore considered in the first two sub-sections in this literature review. The literature review is completed with some discussion of previous work directed at SDG forecasting.

2.1 Short Time Series Forecasting

Short time series forecasting is challenging because it is difficult to perform meaningful out of sample evaluation, or cross validation, given a low number of observations [13]. From the literature a range of methods have been proposed to address this issue, see for example [7]. However, the proposed solutions tend to still insist on 50 or more observations. In the case of the SDG data, the sample size is less than 20 points. The FBProphet time series forecasting tool was used in [2] for the purpose of SDG attainment prediction where it was demonstrated that FBProphet produced a better prediction accuracy over two alternatives, ARMA and ARIMA.

However, FBProhpet is a uni-variate predictor; given that the focus of this paper is prediction using sets of causal-related time series a multi-variate approach is required. A multivariate time series forecasting model, using Long Short Term Memory (LSTM) networks, was presented in [14]. The LSTM model demonstrated a better overall performance compared to ARMA and ARIMA [7]. The LSTM model was adopted in [4] for multi-variate SDG attainment forecasting. More generally, LSTM models have been widely adopted with respect to many real-life applications such as weather [20] and stock market [6] prediction. With respect to the work presented in this paper an Encoder-Decoder LSTM, was used [14]. LSTM typically performs better when large data sets are used. But also seems to perform well when a large number of short time series are available, as in the case of the SDG prediction application considered here.

2.2 Time Series Causal Inference

Causal inference is concerned with the process of establishing a connection (or the lack of a connection) between events or instances. Given two candidate time series, $A = \{a_1, a_s, \ldots, a_n\}$ and $B = \{b_1, b_2, \ldots, b_m\}$, where we wish to establish that B is causality-related to A, this is typically established using a prediction mechanism that uses the "lag" $\{b_1, \ldots, b_{m-1}\}$ to predict a_n. We then compare the predicted value for a_n with the known value, for example using the Root Mean Square Error (RMSE) as a comparison metric. If the two values are close then we can say that the "time series A is causality-related to time series B".

There are a number of mechanisms that can be adopted to achieve the above. With respect to the work presented in this paper, six such mechanisms were considered: (i)

Granger Causality (GC), (ii) the Temporal Causal Discovery Framework (TCDF), (iii) Pearson coefficient, (iv) Lasso, (v) the Mann-Whitney U Test and (vi) ACA. Each is discussed in some further detail below.

Granger Causality. Granger Causality (GC) is one of the most widely used causal inference mechanisms found in the literature [8,17]. It was introduced in the 1960s s and is calculated as shown in Eq. 1 where: (i) X and Y are time series, (ii) a and b are the laggs of X and Y, (iii) t is the current time step and (iv) e is a residual error. The idea is that if time series X "granger causes" time series Y, then the past values of X should contain helpful information to forecast Y in a manner that would be better than when forecasting y using only historical data associated with Y. The variation of GC that was used with respect to the research presented in this paper is the Stats-models variation [23]. GC has been used previously in the context of SDG prediction, for example in [8] 20,000 pairs of time series that featured causal relationship were found.

$$Xt = a_1 X_{t-1} + b_1 Y_{t-1} + e \tag{1}$$

Temporal Causal Discovery Framework. The Temporal Causal Discovery Framework (TCDF) [18] is an alternative mechanism to GC to determine whether a time series A has a causal association with a time series B. TCDF uses a Convolutional Neural Network (CNN) whose internal parameters are interpreted to discover causal relations. The framework has been shown to not work well with respect to short time series. For best performance it is suggested that 1000 data points are required, but is still considered in this paper.

Pearson Correlation. Pearson Correlation [10] has been used to measure the correlations between any given pair of time series. The mechanism assumes linearity of the data. This assumptions holds with respect to many SDG time series that are typically linearly spaced, and therefore seems an appropriate choice.

Lasso. Lasso [26] is an L1 regularisation technique frequently used to reduce high dimensionality data, which can also be employed to establish the existence of a causality between variables [9,26]. LASSO reduces the dimensionality of the input data set by penalising variances to zero, thus allowing irrelevant variables to be removed. Equation 2 shows the LASSO cost function. Inspection of the equation indicates that the first part is the *squared error* function, whilst the second part is a penalty applied to the regression slope. If λ is equal to 0, then the function becomes a normal regression. However, if λ is not 0 coefficients are penalised accordingly, leaving only coefficients that can explain the variance in the data.

$$LCF = \sum_{i=1}^{n} \left(y_i - \sum_j x_{ij}\beta_j \right)^2 + \lambda \sum_{j=1}^{p} |\beta_j| \tag{2}$$

Mann-Whitney U Test. The Mann-Whitney U Test [1] is the fifth causal inference mechanism used in this paper. The test is used to determine if any two pairs of time series are statistically different. It is a non-parametric test (unlike, for example, Lasso).

ACA. The last of the six causality discovery mechanisms considered in this paper is the ACA mechanism proposed in [4]; the name is derived from the author's initials. Essentially this is an ensemble of the above five mechanisms which was found to outperform the above mechanisms when used individually.

2.3 Sustainable Development Goals Forecasting

Previous work directed at the forecasting of SDG attainment can be divided into two main categories: (i) single target forecasting or (ii) multiple target forecasting. The first is directed at forecasting with respect to an individual SDG or specific geographical location. Much existing work falls into this category. Examples can be found in [21] and [24] where forecasting was directed at a specific SDG (electricity supply) and specific region (Ukraine) respectively. The second is concerned with predicting multiple targets. Examples of this second approach include the SDG-AP and SDG-CAP frameworks presented in [2] and [4] respectively that were referenced in the introduction to this paper and that are used for comparison purposes with respect to the evaluation of the SDG TTF framework given later in this paper.

3   The SDG Data Set and Associated Data Preparation

To maintain oversight of the SDG agenda, the UN periodically releases SDG related data on the www platform of the United Nations Department of Economic and Social Affairs' Statistics Division[2]. Once on the SDG data website, the data can be downloaded, partially or wholly, in a CSV format. The SDG platform holds data related to 346 countries. In addition the platform features collated data for regional groupings, such as Sub-Saharan Africa, Northern Africa, Western Asia, Central and Southern Asia, and so on. The total number of indicators is 561 divided across 169 targets (and 17 SDGs). Each indicator will have one or more sub-indicators.

The SDG data set comprises a set of records $\{R_1, R_2, \dots\}$. Each record R_i comprises a set of values $\{v_1, v_2, \dots\}$ where each value corresponds to a set of attributes $A = \{a_1, a_2, \dots\}$. The attributes take either a categorical or numerical value. Thus D comprises a single, very large, table with the columns representing a range of numerical and categorical attributes and the rows representing single observations related to individual SDG indicators. Each record R_i is date stamped. The set A (the columns in the table) represent the complete set of attributes for all 561 indicators. However, for any one indicator only a small sub-set of the available set of attributes will be relevant. Table 3 gives an example of a SDG record for the country Afghanistan for the year 2015. The example refers to Goal 16 "*Promote peaceful and inclusive societies for sustainable development, provide access to justice for all and build effective, accountable and inclusive institutions at all levels*", Target 16.1 is then "*Significantly reduce all forms of violence and related death rates everywhere*", which has associated with it Indicator 16.1.1, "*Number of victims of intentional homicide per 100,000 population, by sex (victims per 100,000 population)*", which in this case has a value of 0.55597 per 100,000

---

[2] https://unstats.un.org/sdgs/indicators/database.

head of population. For ease of reading the table has been arranged in a multi-column format, the record is actually a single row in the SDG database table. From the example it can be seen that many of the attributes are not relevant (such as "Mountain elevation" or "type Of Speed"). This is why the value for many attributes in Table 3 has been set to "NA" (Not Applicable). This is a feature of all the records in the SDG data set.

Note that the indicator reference, 16.1.1 in above the example, incorporates the goal and target references, hence for further processing we only need the indicator reference. In the remainder of this paper we will refer to this as the Goal-Target-Indicator (GTI). Note that targets are also identified, in the raw SDG data, using lower case letters, for example 9.c.1. The GTI is sufficient to identify individual rows if there is only one sub-indicator. However, in many cases we have more than one sub-indicator. Hence, to differentiate between individual time series, a unique "Individual Series" (IS) identifier, made up of the Series Code (the fourth attribute in Table 3, and further identifying characters, was devised. Note, from the table, that the series code is made up of three *text segments* separated by underscore characters. IS values were constructed by adding a fourth text segment.

To allow SDG prediction the SDG data, as described above, needed to be preprocessed into a structured format. In [2] and [4] prediction was facilitated by a hierarchical taxonomy SDGs $\Rightarrow$ Targets $\Rightarrow$ Indicators $\Rightarrow$ Sub-indicators with predictors at the sub-indicator leaf nodes whose results were passed up the tree and combined level-by-level till the root of the tree was reached and a final prediction arrived at. The same approach was adopted with respect to the SDG TTF framework presented here. The structured format thus also had to facilitate the population of the taxonomy, once generated, with respect to individual countries (geographic regions). An important element of populating the taxonomy was the collation of the time series values to be used to create the predictors to be held at the taxonomy leaf nodes. The mechanism adopted with respect to the SDG TTF framework described here for creating the predictors is what sets it apart from the SDG-AP and SDG-CAP frameworks described in [2] and [4].

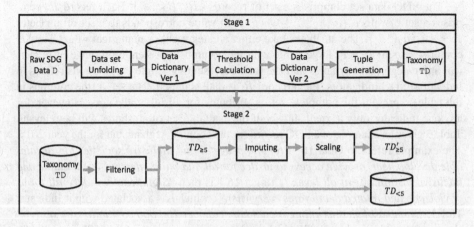

Fig. 2. Preprocessing Schematic.

Table 3. SDG Example Record.

Attribute	Goal	TimeCoverage	Cities
Value	16	NA	NA
Attribute	Target	UpperBound	Counterpart
Value	16.1	NA	NA
Attribute	Indicator	LowerBound	Disability status
Value	16.1.1	NA	NA
Attribute	SeriesCode	BasePeriod	Education level
Value	VC_IHR_PSRC	NA	NA
Attribute	SeriesDescription	Source	Fiscal intervention stage
Value	Number of victims of intentional homicide per 100,000 population, by sex	National Statistical Organization	NA
Attribute	GeoAreaCode	GeoInfoUrl	Food Waste Sector
Value	4	NA	NA
Attribute	GeoAreaName	FootNote	Freq
Value	Afghanistan	NA	NA
Attribute	TimePeriod	Activity	Frequency of Chlorophyll-a concentration
Value	2015	NA	NA
Attribute	Value	Age	Grounds of discrimination
Value	0.55597	NA	NA
Attribute	Time_Detail	Cause of death	Hazard type
Value	2015	NA	NA
Attribute	IHR Capacity	Mode of transportation	Observation Status
Value	NA	NA	NA
Attribute	Level of requirement	Mountain Elevation	Parliamentary committees
Value	NA	NA	NA
Attribute	Level/Status	Name of international institution	Policy Domains
Value	NA	NA	NA
Attribute	Location	Name of non-communicable disease	Policy instruments
Value	NA	NA	NA
Attribute	Nature	Quantile	Migratory status
Value	C	NA	NA
Attribute	Report Ordinal	Substance use disorders	Type of speed
Value	NA	NA	NA
Attribute	Reporting Type	Type of occupation	Type of support
Value	G	NA	NA
Attribute	Sampling Stations	Type of product	Type of waste treatment
Value	NA	NA	NA
Attribute	Sex	Type of skill	Units
Value	FEMALE	NA	PER_100000_POP

A schematic of the adopted pre-processing mechanism is given in Fig. 2. From the figure it can be seen that the mechanism comprised two stages.

Stage 1. Taxonomy Generation.
Stage 2. Missing Value Imputation and Scaling and Generation of Country Data Files.

Each stage is discussed in further detail in the following three subsections.

3.1 Taxonomy Generation (Pre-processing Stage 1)

Stage 1, as noted above, comprises taxonomy generation. The taxonomy, although describing a tree structure, is actually stored as a set of tuples in a data set $TD = \{TR_1, TR_2, \dots\}$, where each $TR_i \in TD$ is a tuple of the form:

$$TR_i = \langle Geographical\_Region, GTI, IS, T \rangle \tag{3}$$

where: (i) "Geographical_Region" is the name of the country or region of interest, (ii) GTI is the relevant Goal-Target-Indicator, (iii) IS is the relevant Individual Series identifier for the time series held at the leaf node in the topology for a given region, and (iv)

T is the time series associated with the leaf node, $T = [t_1, t_2, \ldots]$. Note that the particular relevance of the format of TD is that it allows comparison between records within a given geographic region, and comparison across geographic regions, a central feature of the proposed Extended SDG-TTF framework.

Pre-processing Stage 1 is thus about transforming the raw SDG D input into a data set TD. The process commences by applying a depth-first "unfolding" operation to the raw data D so as to collect the topology paths for particular region and indicator pairs. The path information is stored in a data dictionary, and intermediate data repository between D and TD designed to facilitate the transformation from D to TD. An example of this unfolding, using Afghanistan and GTI 16.1.1, is given in Table 3, is given in Fig. 3.

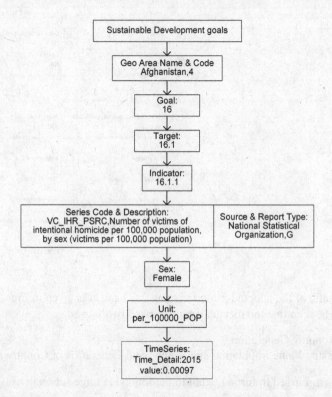

Fig. 3. Illustration of the SDG raw data depth-first "unfolding" for the region-GTI pair Afghanistan and 16.1.1

A number of example dictionary entries are given below:

1. {(Goal : 16), (Target : 16.1), (Indicator : 16.1.1), (Series Code : VC_IHR_PSRC), (Series Description : Number of victims of intentional homicide per 100,000 population , by sex (victims per 100,000 population))) (Geo Area Name : Afghanistan), (Units : PER_100000_POP), (Sex : FEMALE)}
2. {(Goal : 11), (Target : 13.1), (Indicator : 11.b.2), (Series Code : SG_GOV_LOGV), (Series Description : Number of local governments (number)), (Geo Area Name : Afghanistan), (Units : NUMBER)}

3. {*(Goal : 15), (Target : 15.4), (Indicator : 15.4.2), (Series Code : ER_MTN_ GRNCVI),*
   (Series Description : Mountain Green Cover Index), (Geo Area Name : Afghanistan),
   (Units : PERCENT), (Observation Status : A), (Mountain Elevation : 4)}

The first of the above examples is for the region-GTI pair Afghanistan and 16.1.1 used for illustrative purposes in Table 3 and Fig. 3. The second two are two additional examples. Note that only the most salient information is stored in the dictionary, the information needed for the topology.

The next stage is to produce a mathematical formulation of the threshold value to be used to determine whether each indicator has been met or not. In some cases this is straight forward, in others this is not so straight forward. For example phrases such as *"Eradicate"* and *"reduce at least by half"* are used to define TGIs. A solution, in the context of the proposed taxonomy, was available in [16] where the authors published guidelines on how to interpret the health target goals from the SDG published Target Goals document, including mathematical definitions. The guidelines in [16] were adopted to translate the textual descriptions of SDG sub-indicators into mathematical ones. We now have all the information needed for our taxonomy.

Table 4 summarises the details of the sub-indicators held in the dictionary with respect to GTIs 1.1.1, 1.2.1 and 1.3.1. In the table, Column 1 gives the GTI, Column 2 gives the "Thresholds" (calculated as described above) that need to be met before the dead line given in Column 3, Column 4 gives the "Series Description", Column 5 the IS indicator and Column 6 the unique ID number for the leaf node. Recall that the IS (Individual Series) indicator is a textual description of sub-indicator and by extension the associate time series. IS identifiers are constructed from the raw data (see Table 3). As noted earlier, the IS indicator comprises four text segments separated by underscore characters. The first three are the series code taken from the raw data (see Table 3). For example, considering the fifth example in Table 4, "VC_IHR_PSRC". The fourth text segment that has been added in this case is "FEMALE" to give an IS indicator of "VC_IHR_PSRC_FEMALE". The unique ID is simply a unique numeric identifier which is simpler to use than the IS indicator (but not at all easy to interpret).

The final step in stage one (see Fig. 2) is to transpose the data in the dictionary so that it is held in a set of the form $TD = \{TR_1, TR_2, ...\}$ where each $TR_i \in TD$ is a tuple of the form described above. Note that to reference the time series associated with a record TR_i we will use the notation T_i. Table 5 gives a series of example time series for a number of indicators for Afghanistan. Each column represents a time series. The column headings give the relevant GTI and the time series ID. What can be clearly seen from the table is that there are many missing values!

3.2 Missing Value Imputation and Scaling (Pre-processing Stage 2)

The theoretical maximum length of any SDG time series is 22 points, covering 22 years of observations from 2000 to 2021. Some indicators, for a small number of countries, have data going back to 1974. Figure 4 shows the number of observations per year with respect to the 30 different countries considered for evaluation purposes in this paper. Inspection of the figure indicates that the majority of the data falls within 2000 and 2018. Beyond 2018 the data is frequently not yet available (in some cases it may never become

Table 4. Examples taxonomy leaf node information held in the Data Dictionary.

G.T.I	Thresholds	Date	Series Description	Series Code	ID
1.1.1	<=0.5%	2030	Employed population below international poverty line, by sex and age (%)	SI_POV_EMP_15-24_MALE	1
				SI_POV_EMP_BOTHSEX_15+	2
				SI_POV_EMP_BOTHSEX_15-24	3
				SI_POV_EMP_BOTHSEX_25+	4
				SI_POV_EMP_FEMALE_15+	5
				SI_POV_EMP_FEMALE_15-24	6
				SI_POV_EMP_FEMALE_25+	7
				SI_POV_EMP_MALE_15+	8
				SI_POV_EMP_MALE_25+	9
			Proportion of population below international poverty line (%)	SI_POV_DAY	10
1.2.1	<=50%		Proportion of population living below the national poverty line (%)	SI_POV_NAHC__ALLAREA	11
1.3.1	>=80%		Proportion of children/households receiving child/family cash benefit (%)	SI_COV_CHLD_BOTHSEX	12
				SI_COV_CHLD_FEMALE	13
				SI_COV_CHLD_MALE	14
			[ILO] Proportion of employed population covered in the event of work injury (%)	SI_COV_WKINJRY_BOTHSEX	15
				SI_COV_WKINJRY_FEMALE	16
				SI_COV_WKINJRY_MALE	17
			[ILO] Proportion of mothers with new borns receiving maternity cash benefit (%)	SI_COV_MATNL_BOTHSEX	18
				SI_COV_MATNL_FEMALE	19
			[ILO] Proportion of poor population receiving social assistance cash benefit (%)	SI_COV_POOR_BOTHSEX	20
			[ILO] Proportion of population above statutory pensionable age receiving a pension, by sex (%)	SI_COV_PENSN_BOTHSEX	21
				SI_COV_PENSN_FEMALE	22
				SI_COV_PENSN_MALE	23
			[ILO] Proportion of population covered by at least one social protection benefit (%)	SI_COV_BENFTS_BOTHSEX	24
				SI_COV_BENFTS_FEMALE	25
				SI_COV_BENFTS_MALE	26
			[ILO] Proportion of population with severe disabilities receiving disability cash benefit (%)	SI_COV_DISAB_BOTHSEX	27
				SI_COV_DISAB_FEMALE	28
				SI_COV_DISAB_MALE	29
			[ILO] Proportion of unemployed persons receiving unemployment cash benefit, by sex (%)	SI_COV_UEMP_BOTHSEX	30
				SI_COV_UEMP_FEMALE	31
				SI_COV_UEMP_MALE	32
			[ILO] Proportion of vulnerable population receiving social assistance cash benefit (%)	SI_COV_VULN_BOTHSEX	33
				SI_COV_VULN_FEMALE	34
				SI_COV_VULN_MALE	35
			[World Bank] Poorest quintile covered by labour market programs (%)	SI_COV_LMKTPQ_	36
			[World Bank] Poorest quintile covered by social assistance programs (%)	SI_COV_SOCASTPQ_	37
				SI_COV_SOCINSPQ_	38
			[World Bank] Proportion of population covered by labour market programs (%)	SI_COV_LMKT_	39
			[World Bank] Proportion of population covered by social assistance programs (%)	SI_COV_SOCAST_	40
				SI_COV_SOCINS_	41

available). Thus, in the context of the research presented in this paper, only data from 2000 to 2018 was considered; 34, 526 time series in total. Therefore, with reference to Fig. 2, the first step in Stage 2 was to filter TD so that it comprised only of 18 point time series.

As noted with reference to Table 5, the SDG data set features a significant number of missing values.

The reasons for missing data in the collated SDG time series are varied but can be categorised as either: Missing At Random (MAR) or Not Missing At Random (NMAR) [12]. We can illustrate the distinction by considering the thwo example time series given in Table 6, the first describes the time series for the indicator *"Direct economic loss attributed to disasters (current United States dollars)"* (GTI 11.5.2, ID 3109). Inspection of the associated time series reveals a time series with only one recorded value, the value of 311. However, the data describes *"loss attributed to disasters"*, which by definition (we hope) are not regular occurrences, hence financial losses as a result of disasters are

Table 5. Example time series for Afghanistan and a number of indicators arranged in columns.

Years	1.2.1 (11)	1.4.1 (2269)	1.4.1 (2271)	1.4.1 (2270)	1.5.1 (2957)	8.4.2 (1071)	8.4.2 (1073)	8.4.2 (1087)	8.4.2 (1076)
2000		22.74099	31.29478	24.64515		21000	6945939	1925978	200178
...									
2005	33.7	27.47609	40.17754	30.41177		243004	9645728	2188895	113076
...									
2009	38.3	31.29914	48.66158	35.4571		725012	9408891	2350763	113076
...									
2015		37.0523	62.26144	43.41761	17	1820623	12948523	2395294	107546
...									
2018									

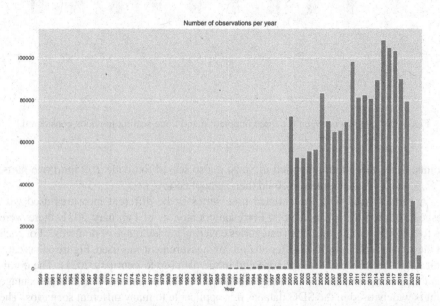

Fig. 4. Number of Observation per year with respect to the 30 different countries considered for evaluation purposes in this paper.

not recorded every year. This type of missing data is thus considered to be NMAR data. The second column describes the time series indicator "*Proportion of population covered by a mobile network, by technology*" (GTI 9.c.1, ID 2712). In this case the absence of the missing data is unclear because Egypt did have mobile services prior to 2014. Thus this type of missing data is considered to be MAR data.

To address the missing data problem the idea was to adopt some kind of data imputation, the process of assigning values to missing attribute value instances according to neighbouring values. This will only work if sufficient neighbouring values are available. Some preliminary experiments, not reported here, indicated that for the imputation to have a chance of success a minimum of 25% of the values were required. In other words, given that our time series were of length 18, we needed values for five or more of the

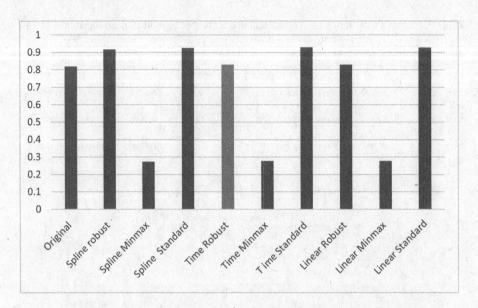

Fig. 5. Comparison between the three imputation and three scaling methods considered.

points. Thus, the filtering applied in Stage 2 also served to divide TD in to two parts, $TD_{<5}$ and $TD_{\geq 5}$. Imputation could then be applied to $TD_{\geq 5}$.

A further issue with the collated time series is the different measures used with respect to the different indicators. For each country, as of February 2021, there were up to 3,408 different time series categories covering a wide range of domains[3]. For each of these time series one of 45 different units of measurement was used. Figure 6 lists each unit and the number of times it appeared in the data (up to February 2021). The dominant measuring unit is the percentage, followed by Tonnes and number. The percentage unit is widely used in the SDG data, as it is applicable to many different scenarios. The Tonnes unit of measurement is used most frequently with respect to Goal 8 "*Promote sustained, inclusive and sustainable economic growth, full and productive employment decent work for all*" where many sub-indicators measure material consumed in a country. The number unit measurement was often used to describe a monetary figure or a population. With this in mind, any consideration on building multivariate time series with the help of causal inference will require all the data to be on the same scale. Without scaling the time series, a series of population counts or of monetary value will always dominate over (say) series comprised of percentages values. There was thus also a requirement for some from of scaling to be applied to the data.

A set of experiments was conducted to find the best mechanism for imputing missing values and for scaling the data in TD. For the experimentation data for 41 countries was used together with three different imputation methods and three different scaling

---

[3] Note that all sub-indicators are not necessarily relevant to all countries, for example sub-indicators concerned with forestation will not be relevant to a desert country, hence all countries do not feature exactly the same number of time series.

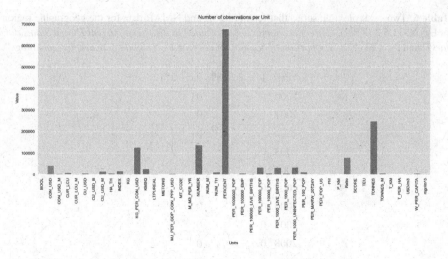

Fig. 6. Number of observations per unit.

algorithms were considered. The imputation method were: (i) Spline, (ii) Time and (iii) Linear [11, 15, 29]. The scaling algorithms considered were: (i) Robust, (ii) Minmax and (iii) Standard [19]. These methods and algorithms were chosen because of their popularity in the literature. The three imputation methods and three scaling algorithms could be combined in nine different ways. Only complete time series where used for the experimentation; time series where all 18 values were available. There were 218 of these out of the 36,742 time series associated with the 30 countries used as a focus for the experimentation. Each time series was split into a training part and a testing part, $T_{i_{train}}$ and $T_{i_{test}}$, four values were then removed, at random, from each $T_{i_{train}}$ and then the selected imputation method and scaling were algorithm applied to $T_{i_{train}}$. The result was then used to predict the values in the test part. The imputation method and scaling algorithm that was the closest prediction match would then be used in Stage 2 of the SDG data preprocessing. For the prediction FBProhpet [25] was used. The adopted evaluation metric was average Root Mean Square Error (RMSE) was used. Although normally a lower RMSE values means a better results, in this experiment the goal was to stay as close as possible to the baseline RMSE value. The results of the experiments are given in Fig. 5 which shows the average RMSE per combination. From the table it can be seen that the Time imputation coupled with Robust scaling (Time&Roubust) produced an average RMSE of 0.8296 which is the closest average RMSE values to the original value of 0.8185. This was then the combination used in Stage 2 to address the missing value and many measurement units used problem.

Returning to Fig. 2, augmentation and scaling was applied to $TD_{\geq 5}$ to give $TD'_{\geq 5}$. However, note that $TD'_{<5}$ was not thrown away. The reason for the later will become clear later in this paper.

Table 6. Two example time series that feature MAR and NMAR data for the geographic region of Egypt; 11.5.2 (3189), *"Direct economic loss attributed to disasters (current United States dollars)"* and 9.c.1 (2712), *"Proportion of population covered by a mobile network, by technology"*.

Year	11.5.2 3189	9.c.1 2712
2000		0
2001		0
2002		0
2003	0	0
2004	0	0
2005	0	0
2006	0	0
2007	0	0
2008	0	0
2009	0	0
2010	0	0
2011	0	0
2012	0	0
2013	0	0
2014	0	61
2015	0	89
2016	311	0
2017	0	95.1

4   The Extended SDG Track, Trace and Forecast (SDG-TTF) Model

This section presents the SDG-TTF framework. The workflow for the framework is presented in Fig. 7. The input is the set of time series, $\mathbf{T} = \{T_1, T_2, \dots\}$ extracted from $TD'_{\geq 5}$ (generated as described in the previous section).

From the figure it can be seen that the extended SDG-TTF framework comprises five processes: (i) Data Grouping, (ii) Relation Discovery, (iii) multivariate Forecasting, (iv) univariate forecasting and (v) bottom-up classification. Note that two forecasting processes, multivariate and univariate, feed into the bottom up classification. The end result is a set of probabilistic SDG attainment predictions for the input set of countries. Each of the five stages will be considered in further detail in the remainder of this section.

During the data grouping process $\mathbf{T}$ is grouped into geographic regions. Recall that the objective of this paper is to improve on current SDG prediction effectiveness by taking into consideration both intra- and inter-causalities, causalities within individual countries and causalities between countries and their neighbours. Something not considered in previous work. The data grouping was conducted using geographic area codes based on the UN regional segmentation[4]. The intuition here was

---

[4] https://unstats.un.org/sdgs/report/2019/regional-groups/.

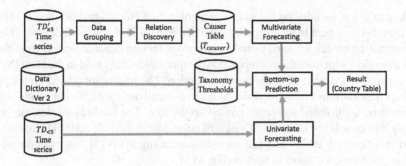

Fig. 7. Overview of the SDG-TTF workflow.

that the SDG performance of neighbouring countries will have an impact on the SDG performance of the country under consideration. At the end of the grouping stage the set $\mathbf{T}$ will have been divide into a set of Region Time series, $\mathbf{T} = \{TR_1, TR_2, \dots\}$

The next process is to determine the causal relationships between the time series in each region time series set TR_i for each grouping. Each $T_i \in TR_i$ is compared to each other time series in TR_i; in other words the complement of $T_i \in TR_i$ denoted as T_i^c. The interaction between each time series is measured using a causality ranking measure r. This is calculated using Root Mean Square Error (RMSE) in a similar manner as described for the experiments described in Sub-sect. 3.2. For the evaluation presented in the following section the six time series causality mechanisms listed in Sect. 2 were used (Lasso, R^2, Pearson Correlation, Mann-Whitney U Test, Granger and ACA). For each T_i, the time series in T_i^c were ranked according to r and the top k selected as having a causal relationship with T_i, the set $T_{i_k}^c$. For the evaluation presented later in this paper $k = 50$ was used. Each T_i and $T_{i_k}^c$ pairing was then stored in a "causer table", $T_{causer} = \{\tau_1, \tau_2, \dots\}$, where $\tau_i = T_i \cup T_{i_k}^c$.

For each $\tau_i \in T_{causer}$ the next process in the workflow, shown Fig. 7, was to build a multi-variate time series forecasting model. Recall that these forecasting models sit at the leaf nodes of the SDG taxonomy populated with respect to a particular country. A range of tools and techniques are available whereby such a model can be constructed. However, for the evaluation presented later in this paper a multi-variate LSTM-Encoder-Decoder (Enc-Dec) [14] was used.

Recall, from the previous section, that during data preprocessing time series which were deemed unusable with respect to the determination of causality relationships were set aside in the set $T_{<5} = \{T_1, T_2, \dots\}$. However, although unsuited to causality relationship determination this data can still be used for the purpose of forecasting SDG attainment. For each time series $T_i \in T_{<5e}$ a uni-variate time series forecasting model was built (to be held at the relevant leaf nodes within the taxonomy). Again there are a number of tools and techniques available whereby such a model can be constructed. For the evaluation presented in the following section uni-variate FBProphet was used.

The final process in the extended SDG-TTF workflow (Fig. 7) is the prediction process where we ascertain whether a given country will meet its SDG goals or not using the generated multi-variate and uni-variate time series forecasting models described above. The fundamental process is similar to that of the SDG-AP framework presented in [2], which in turn was founded on the same hierarchical SDG topology described in [4] and used again with respect to the proposed extended SDG-TTF framework described here. The forecasting models are used to make predictions for individual indicators (leaf nodes in the topology) which are then compared to the threshold values held in the Data Dictionary generated as described in the previous section. The results are passed up the SDG topology hierarchy up to the root node. At each intermediate node a Boolean "and" operation will be applied and the result passed up the tree. The final result, that the given country will or will not attain its SDGs, will culminate at the root node. The results are stored in a "country table" and can be visualised using D3.js [5]. An example of the latter is given and discussed in Sect. 6 (Fig. 8).

5   Evaluation

The evaluation of the proposed extended SDG-TTF model is presented in this section. For the evaluation the UN North Africa, South Asia and Northern Europe geographic regions, as defined by the UN Geoscheme, were considered:

North Africa. Algeria, Egypt, Libya, Morocco, Sudan, Tunisia and Western Sahara.
South Asia. Afghanistan, Bangladesh, Bhutan, India, Iran, Maldives, Nepal, Pakistan and Sri Lanka.
Northern Europe. Aland Islands, Denmark, Estonia, Faroe Islands, Finland, Iceland, Ireland, Isle of Man, Latvia, Lithuania, Norway, Svalbard and Jan Mayen Islands, Sweden and United Kingdom.

This comprised a total of 23,068 time series (leaf nodes in the topology), covering the 17 SDGs with respect to the sub-region of interest. After pre-processing (see Sect. 3) $TD'_{\geq 5}$ comprised 8,629 and $TD_{<5}$ 15,439 time series. The substantial number of time series allocated to $TD_{<5}$ was due to the large number of missing values that featured in the data.

The objectives of the evaluation were:

1. To determine the most appropriate causality discovery mechanism for use with the SDG-TFF framework.
2. To determine whether, by taking into consideration both intra-region and inter-region causality relationships, better SDG predictions could be produced.

For the evaluation the each time series was divided into two arts, the first 14 observations were used for training, and the last 4 observations for testing; $k = 50$ was used through out. All experiments were run on a windows 10 machine running under Ryzen 9 CPU, RTX 2060 GPU, 40 GB of RAM and 1 TB SSD. Comparisons were made with the SDG-AP and SDG-CAP prediction frameworks presented in [2] and [4] respectively. Recall that using SDG-CAP only intra-entity (single country) causal relationships were considered, as opposed inter-entity causal relationships as in the case of SDG-TTF. For SDG-AP framework two prediction models were considered, LSTM and FBProphet.

Table 7. A sample of RMSE values for selected SDG indicators for Afghanistan.

Time Series Code		SDG-TTF						SDG-CAP	SDG-AP	
		Lasso	R2	pearson	T_test	Granger causality	ACA	ACA	LSTM	FBProphet
1	afghanistan_1.4.1_ 2269	0.645	0.544	0.589	1.000	0.5585	0.6028	0.6910	0.313	0.0008
2	afghanistan_1.4.1_ 2270	0.763	0.672	0.664	0.592	1.0000	0.6253	0.6337	0.310	0.0086
3	afghanistan_1.4.1_ 2271	0.664	0.690	1.000	0.692	0.6158	0.7452	0.4122	0.297	0.0159
4	afghanistan_1.4.1_ 2272	1.000	0.675	0.662	0.680	0.6749	0.6022	0.5578	0.308	0.0150
5	afghanistan_1.4.1_ 2273	0.633	1.000	0.596	0.697	0.5002	0.5771	0.7197	0.288	0.0200
6	afghanistan_1.4.1_ 2274	0.633	0.646	0.668	0.738	1.0000	0.6796	0.6149	0.312	0.0212
7	afghanistan_1.a.1_ 2956	0.191	0.259	0.268	1.000	0.2522	0.2416	0.9103	0.985	0.5219
8	afghanistan_1.a.2_ 2277	0.874	0.868	1.000	0.906	0.8732	0.9049	0.9575	3.304	1.1733
9	afghanistan_10.4.1_ 2721	1.000	0.296	0.308	0.300	0.3159	0.3216	0.3675	1.308	0.0725
10	afghanistan_10.5.1_ 2725	0.251	0.277	1.000	0.261	0.2649	0.2505	0.2716	0.298	2.0310
11	afghanistan_10.5.1_ 2726	0.047	1.000	0.016	0.016	0.0136	0.0239	0.0179	0.283	3.5772
12	afghanistan_10.5.1_ 2727	1.000	0.007	0.056	0.130	0.0575	0.0854	0.1129	0.160	5.8625
Average		0.642	0.578	0.569	0.584	0.510	0.471	0.522	0.680	1.110
Standard Deviation		0.311	0.299	0.330	0.318	0.325	0.265	0.281	0.856	1.7833

All algorithms were implemented using the Python programming language. The evaluation metric used was RMSE (Root Mean Squared Error). As noted earlier, six different causality discovery mechanisms were considered: Lasso, R^2, Pearson Correlation, Mann-Whitney U Test, Granger and ACA.

A sample of the recorded RMSE values for the country Afghanistan and 12 selected SDGs is given in Table 7. The first two columns give the time series ID number and the Individual Series (IS) indicator. The next six columns give the RMSE results obtained using the six causality mechanisms considered and the extended SDG-TTF framework. The following column gives the results obtained using SDG-CAP and ACA causality as described in [4] and the last two columns the results obtained using SDG-AP coupled with Univariate LSTM and FBProphet as described in [2]. The average RMSE value is given at the bottom of the table, for each approach considered, together with the associated standard deviation. From the sample it can be seen that the Extended SDG-TTF framework, coupled with ACA causality, produced the best overall result (highlighted in bold font). These results were confirmed by inspection of the complete set of results (not shown here) for the three regions considered. It was thus concluded that the most appropriate causality discovery mechanism was the ACA mechanism.

Tables 8, 9 and 10 present a summary of the results obtained for the North Africa, South Asia and Northern Europe regions considered, suing: SDG-TTG and ACA causality, SDG-CAP and ACA causality, SDG-AP with LSTM and SDG-AP with FBProphet. From tables it is clear that consideration of inter-entity causal relationships, as well as intra-entity causal relationships, as incorporated into the SDG-TTF framework, results in an improved SDG attainment prediction.

Table 8. Average RMSE values for the North Africa geographic region per country [3].

Country	SDG-TTF (ACA)		SDG-CAP (ACA)		SDG-AP (FBProphet)	
RMSE	AVG	SD	AVG	SD	AVG	SD
Algeria	0.3	0.5	0.4	0.9	0.8	7.6
Egypt	0.4	1.4	0.5	2.0	0.6	3.1
Libya	0.8	1.1	0.9	1.0	0.6	0.8
Morocco	0.6	0.3	0.5	1.4	0.6	1.3
Sudan	0.2	0.2	0.3	0.3	0.4	0.4
Tunisia	0.4	0.8	0.5	1.1	0.7	1.8
Western Sahara	0.5	0.3	0.6	0.5	0.8	0.5
Average	0.4	0.7	0.5	1.0	0.6	2.2

Table 9. Average RMSE values for the South Asia geographic region per country [3].

Country	SDG-TTF (ACA)		SDG-CAP (ACA)		SDG-AP (FBProphet)	
RMSE	AVG	SD	AVG	SD	AVG	SD
Afghanistan	1.305	7.228	1.310	7.231	1.887	2.424
Bangladesh	1.005	3.751	1.089	2.614	1.780	2.611
Bhutan	1.481	8.055	1.452	4.354	3.352	40.989
India	1.480	8.079	0.744	0.812	0.984	2.481
Iran	1.321	7.666	0.760	1.194	1.938	19.521
Maldives	1.298	1.285	3.624	23.232	0.991	3.256
Nepal	7.658	7.230	1.267	1.643	1.518	3.844
Pakistan	0.490	0.507	0.930	1.672	3.754	65.313
Sri Lanka	0.436	0.520	1.181	2.230	1.958	2.964

6 System Operation

The operation of the SDG-TTF framework was investigated using a number of case studies. One such case study is partly presented here. Namely, SDG 3, Target 2: *"By 2030, end preventable deaths of newborns and children under five years of age, with all countries aiming to reduce neonatal mortality to at least as low as 12 per 1000 live births and under-5 mortality to at least as low as 25 per 1000 live births"*, and the country Algeria. Target 3.2 thus comprises two indicators (3.2.1 and 3.2.2), *"Under-five mortality rate"* and *"Neonatal mortality rate"*. Note that neonatal interpreted as aged less over 1 month old. For the first we have four sub-indicators (time series): (i) deaths of female children aged less than one (1Y/F), (ii) deaths of male children aged less than one (1Y/M), (iii) deaths of female children aged less than five (5Y/F) and (iv) deaths of male children aged less than five (5Y/M). The threshold in this case is $\leq 25\ per$ 1000. For the second we have one sub-indicator (time series): deaths of children under one month of age (1Month/FM) for which the threshold is $\leq 12\ per$ 1000.

Table 10. Average RMSE values for the Northern Europe geographic region per country.

Country	SDG-TTF (ACA)		SDG-CAP (ACA)		**SDG-AP** (FBProphet)	
RMSE	AVG	SD	AVG	SD	AVG	SD
Aland Islands	0.187	0.125	0.197	0.149	0.227	0.189
Denmark	0.637	0.724	4.421	70.753	2.249	24.416
Estonia	0.644	0.726	1.705	3.792	3.567	31.441
Faroe Islands	0.647	0.724	0.877	1.100	0.381	0.693
Finland	0.644	0.722	1.580	3.562	0.813	4.094
Iceland	0.644	0.728	0.360	0.463	23.875	317.859
Ireland	0.608	0.711	0.682	0.158	261.329	4989.750
Isle of Man	0.607	0.707	0.572	0.480	0.156	0.316
Latvia	0.348	0.303	0.592	0.352	1.010	3.822
Lithuania	0.348	0.444	0.585	0.304	26.780	486.870
Norway	1.675	1.537	0.219	0.272	0.728	2.488
Svalbard and Jan Mayen Islands	0.401	0.0274	0.388	0.248	0.0708	0.577
Sweden	0.313	0.355	0.398	0.479	2.574	15.406
United Kingdom	0.336	0.355	1.080	1.714	1799.020	37049.630

Table 11. Forecast results for Target 3.2, for the target year 2030, and the country Algeria [3].

GTI	Age/Sex	Initial	Target	Forecast	Result
3.2.1	1Y/F	20.2	<=25	16.94	Met
3.2.1	1Y/M	22.9	<=25	20.82	Met
3.2.1	5Y/F	23.7	<=25	19.89	Met
3.2.1	5Y/M	26.6	<=25	24.13	Met
3.2.2	1Month/FM	15	<=12	13.75	Not Met

SDG-TTF was then used to make predictions for the year 2030. The generated output is a "country table", as indicated in the workflow presented in Fig. 7. A fragment of this table for Target 3.2 is given in Table 11. The first column gives the GTI. The second gives the sub-indicator (as described above). The third gives the gives the mortality value per 1000 live births in 2015 which is the base year for SDGs. The fourth gives the target thresholds for TGI 3.2.1 and 3.2.2, $\leq 25\ per\ 1000$ and $\leq 12\ per\ 1000$ respectively. The fifth column gives the predicted mortality value per 1000 live births for 2030. The sixth column gives the binary classification, "Met" or "Not Met". For Target 3.2 to be attained (met), the predicted value for each sub-indicator must meet its threshold (at or below the relevant threshold value in this case). In this partial example, all of the included sub-indicators for indicators GTI 3.2.1 are met, unfortunately GTI 3.2.2 is not met.

The software for the Extended SDG-TTF framework includes a visualisation mechanism, as indicated in Fig. 7. This was implemented using D3.js [5]. The visualisation allows users to: (i) track the progress of different goals over a given time frame, and

(ii) trace the achievement of individual bottom level indicators in an interactive manner. An example of such a visualisation is given in Fig. 8 using the case study presented above. From the figure it can be seen that using the visualisation it is easy to identify goal attainment (or non-attainment as in this case). Nodes coloured in green highlight goals, targets, indicators and sub-indicators that will be attained on time. Nodes coloured in red highlight goals, targets, indicators and sub-indicators that will not be attained on time. For a more detailed analysis of why a goal is not attaining the relevant country table can be inspected.

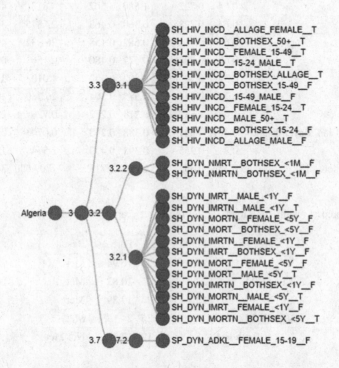

Fig. 8. Visualising of SDG attainment for part of goal 3 [3].

7   Conclusion

In this paper, we have presented an extended analysis of the SDG-TTF attainment prediction framework [3], which, unlike previous frameworks directed at SDG attainment prediction, considers inter- and intra-geographic entity (county, region) causal relationships. It is argued in this paper that individual SDG sub-indicators should not be considered in isolation, in other words in terms of an individual time series, because inspection of the indicators demonstrates clear potential for causal relations with other indicators for a given geographic entity, and potential for causal relationships with the indicators for neighbouring geographic region. The evaluation of the framework shows that more accurate SDG attainment predictions using the SDG-TTF framework can be made. For future work, the authors intend to expand the investigation using more data sources than simply the SDGs data, and consider using alternative causal relationship discovery mechanisms. Finally, the authors intend to evaluate the effect of natural disasters, such as the COVID-19 pandemic, which occur for short periods, on SDG attainment prediction. Note that all the data provided in this paper can be found in the project Github repository[5].

References

1. Alam, N., Rudin, C.: Robust Nonparametric Testing for Causal Inference in Observational Studies, pp. 1–39. Optimization Online, December 2015
2. Alharbi, Y., Arribas-Bel, D., Coenen, F.: Sustainable development goal attainment prediction: a hierarchical framework using time series modelling. In: IC3K 2019, vol. 1, pp. 297–304 (2019). https://doi.org/10.5220/0008067202970304
3. Alharbi, Y., Arribas-Bel, D., Coenen, F.: Sustainable development goals monitoring and forecasting using time series analysis. In: Fred, A.L.N., Sansone, C., Madani, K. (eds.) Proceedings of the 2nd International Conference on Deep Learning Theory and Applications, DeLTA 2021, Online Streaming, 7–9 July 2021, pp. 123–131. SCITEPRESS (2021). https://doi.org/10.5220/0010546101230131
4. Alharbi, Y., Coenen, F., Arribas-Bel, D.: Sustainable development goal relational modelling: introducing the SDG-CAP methodology. In: Song, M., Song, I.-Y., Kotsis, G., Tjoa, A.M., Khalil, I. (eds.) DaWaK 2020. LNCS, vol. 12393, pp. 183–196. Springer, Cham (2020). https://doi.org/10.1007/978-3-030-59065-9_15
5. Bostock, M., Ogievetsky, V., Heer, J.: D3 data-driven documents. In: IEEE TVCG 17, December 2011. https://doi.org/10.1109/TVCG.2011.185
6. Chen, K., Zhou, Y., Dai, F.: A LSTM-based method for stock returns prediction: a case study of China stock market. In: IEEE Big Data 2015. IEEE (2015). https://doi.org/10.1109/BigData.2015.7364089
7. De Gooijer, J.G., Hyndman, R.J.: 25 years of time series forecasting. IFJ 22(3), 443–473 (2006). https://doi.org/10.1016/j.ijforecast.2006.01.001. http://www.sciencedirect.com/science/article/pii/S0169207006000021
8. Dörgo, G., Sebestyén, V., Abonyi, J.: Evaluating the interconnectedness of the sustainable development goals based on the causality analysis of sustainability indicators. Sustainability (Switzerland) 10(10), 3766 (2018). https://doi.org/10.3390/su10103766
9. Epprecht, C., Guegan, D., Veiga, Á.: Comparing variable selection techniques for linear regression: LASSO and Autometrics. Centre d'économie de la Sorbonne (2013). http://halshs.archives-ouvertes.fr/halshs-00917797/
10. Frey, B.B.: Pearson correlation coefficient. In: The SAGE Encyclopedia of Educational Research, Measurement, and Evaluation, pp. 1–4 (2018). https://doi.org/10.4135/9781506326139.n510
11. Hall, C.A., Meyer, W.W.: Optimal error bounds for cubic spline interpolation. J. Approx. Theor. 16(2), 105–122 (1976). https://doi.org/10.1016/0021-9045(76)90040-X
12. Heitjan, D.F., Basu, S.: Distinguishing "missing at random" and "missing completely at random". Am. Stat. 50(3), 207–213 (1996)
13. Hyndman, R., Kostenko, A.: Minimum sample size requirements for seasonal forecasting models. Foresight 6(Spring), 12–15 (2007)
14. Jason, B.: Deep Learning For Time Series Forecasting, vol. 1. Machine Learning Mastery (2018)

---

[5] https://github.com/Yassir-Alharbi/Sustainable-Development-goals.

15. Junninen, H., Niska, H., Tuppurainen, K., Ruuskanen, J., Kolehmainen, M.: Methods for imputation of missing values in air quality data sets. Atmos. Environ. **38**(18), 2895–2907 (2004). https://doi.org/10.1016/j.atmosenv.2004.02.026

16. Lozano, C.J.: Measuring progress from 1990 to 2017 and projecting attainment to 2030 of the health-related Sustainable Development Goals for 195 countries and territories: a systematic analysis for the Global Burden of Disease Study 2017. Lancet **392**(10159), 2091–2138 (2018). https://doi.org/10.1016/S0140-6736(18)32281-5

17. Narayan, P.K., Smyth, R.: Multivariate granger causality between electricity consumption, exports and GDP: evidence from a panel of Middle Eastern countries. Energy Policy **37**(1), 229–236 (2009). https://doi.org/10.1016/j.enpol.2008.08.020

18. Nauta, M., Bucur, D., Seifert, C.: Causal discovery with attention-based convolutional neural networks. Mach. Learn. Knowl. Extr. **1**(1). https://doi.org/10.3390/make1010019

19. Pedregosa, F., et al.: Scikit-learn: machine learning in Python. J. Mach. Learn. Res. **12**, 2825–2830 (2011)

20. Qing, X., Niu, Y.: Hourly day-ahead solar irradiance prediction using weather forecasts by LSTM. Energy **148**, 461–468 (2018). https://doi.org/10.1016/j.energy.2018.01.177

21. Rivera-González, L., Bolonio, D., Mazadiego, L.F., Valencia-Chapi, R.: Long-term electricity supply and demand forecast (2018–2040): a LEAP model application towards a sustainable power generation system in Ecuador. Sustainability (Switzerland) **11**(19), 5316 (2019). https://doi.org/10.3390/su11195316

22. Rockström, J.: Johan Rockström and Pavan Sukhdev present new way of viewing the sustainable development goals and how they are all linked to food. Stockholm Resilience Centre (2016)

23. Seabold, S., Perktold, J.: Statsmodels: Econometric and statistical modeling with Python. In: the 9th Python in Science Conference, pp. 92–96 (2010). https://doi.org/10.25080/majora-92bf1922-011

24. Tkachenko, A., et al.: Efficiency forecasting for municipal solid waste recycling in the context on sustainable development of economy. E3S Web Conf. **166**, 13021 (2020). https://doi.org/10.1051/e3sconf/202016613021

25. Taylor, S.J., Letham, B.: Forecasting at scale. Am. Stat. **72**(1), 37–45 (2018). https://doi.org/10.1080/00031305.2017.1380080

26. Tibshirani, R.: Regression shrinkage and selection via the Lasso. JRSS **58**, 267–288 (1996). https://doi.org/10.1111/j.2517-6161.1996.tb02080.x. http://www.jstor.org/stable/2346178

27. United Nations Statistics Division: E-Handbook on Sustainable Development Goals Indicators

28. United Nations: The Millennium Development Goals Report. United Nations, p. 72 (2015). 978-92-1-101320-7

29. Virtanen, P., et al.: SciPy 1.0: fundamental algorithms for scientific computing in Python. Nat. Met. **17**, 261–272 (2020). https://doi.org/10.1038/s41592-019-0686-2

Disrupting Active Directory Attacks with Deep Learning for Organic Honeyuser Placement

Ondrej Lukas[✉][iD] and Sebastian Garcia[iD]

Faculty of Electrical Engineering, Czech Technical University in Prague,
Prague, Czech Republic
{lukasond,garciseb}@fel.cvut.cz

Abstract. Honeypots have been a long-established form of passive defense in a wide variety of systems. They are often used for the reliability and low false positive rate. However, the deployment of honeypots in the Active Directory (AD) systems is still limited. Intrusion detection in AD systems is a difficult task due to the complexity of the system and its design, where any authenticated account is able to query other entities in the system. Therefore, the positioning of the honeypot in such structures brings two main con trains: (i) the placement has to be organic, with similar properties to other, real entities in the structure, and (ii) the placement must not give away the nature of the honeypot to the attacker. In this work, we present a model based on a variational autoencoder capable of producing organic placements for AD structures. We show that the proposed model is capable of learning meaningful latent representations of the nodes in the AD structures and predicting new node placement with similar properties. Analysis of the latent space shows that the model can capture complex relationships between nodes with low-dimensional latent space. Our method is evaluated based on the (i) similarity with the input graphs, (ii) properties of the generated nodes, and (iii) comparison with other generative graph models. Further experiments with human attackers show that the proposed method outperforms the random honeypot placement baseline.

Keywords: Generative models · Autoencoders · Active directory · Honeypots · Deep learning

1 Introduction

From the wide range of attacks that organizations face, the most critical are those where attackers access the internal network. Large companies such as Sony, Austria Telekom, NTT, and Citrix have been compromised with attackers gaining access to their internal networks [8,9,50,53]. In these attacks, it is usually the Active Directory (AD) that is compromised to gain access to internal resources, user credentials, and sensitive data [11]. AD is a critical part of large organizations and attackers often target it. However, detecting attackers in AD is difficult given that it is used by all members of the organization. It is also hard to distinguish the access of an attacker imposing as a user.

© The Author(s), under exclusive license to Springer Nature Switzerland AG 2023
A. Fred et al. (Eds.): DeLTA 2020/DeLTA 2021, CCIS 1854, pp. 111–133, 2023.
https://doi.org/10.1007/978-3-031-37320-6_6

There are three main ways of defending attackers in an AD environment. First, to stop attackers from *accessing and exploring* the AD. This can be done by using network segmentation and limiting access to critical servers [37], by hardening AD configurations, and by monitoring system events [37,39]. However, this can be a difficult task since users in the domain are supposed to login and use it. Second, to defend an AD is to detect anomalies in the use of AD. Due to the nature of AD, attackers only have to gain access to *any unprivileged* account in *any computer* in the domain to be able to read the majority of AD data [26].

The third way to detect attacks in AD is to use honeyusers. A honeyuser is a fake user account that is disguised as a real user and is designed to attract the attention of the attacker [2]. Since benign users should never interact with the honeyuser, *any interaction* assures the detection of the attacker. Honeyusers have been used for other security detections in the past, such as fake bank accounts and fake database entries [20], but their usage in AD was not explored so far.

The main problem with existing solutions that create honeyusers in AD is that they require the manual intervention of the administrator. Therefore, it is up to them to decide the position and the features of honeyusers. However, to maximize the chance of being selected by an attacker, the placement of the honeyuser is essential. If the honeyuser is inserted in a group in the AD that is not subject to attack, it will probably not be targeted. Another important limitation of manual approaches is the lack of validation, leaving no way of knowing if the honeyuser attracts more attacks.

We propose a deep learning method to automatically generate honeyusers in AD structures, including their features and position in the graph. Our method first captures the characteristics of current AD structures (features of nodes and relationships among nodes) by learning an embedding. Second, it implements a Bidirectional Directed Acyclic Graph Recurrent Neural Network (DAG-RNN), which processes the embeddings together with the matrix of edges of the AD, and to learn the topological relationship of the nodes conditional on their types. The DGA-RNN layer is connected to two Multilayer Perceptrons (MLP) to generate the parameters of a probability distribution that represents the latent space of nodes. The DGA-RNN acts as an encoder for nodes in the graph. The decoder part samples nodes from the distribution and uses a MLP to decide if a node should be connected to each of the previous nodes. At the end, the model outputs a set of new nodes, their features, and *where* they should be connected in the graph in an *organic* way. Finally, our method enriches the honeyusers with attributes and injects them in the original AD structure.

Training a deep learning model needs large amounts of data. However, AD has sensitive information about organizations, and it is extremely difficult to obtain. Therefore, after accessing a small sample of actual AD structures, we used them to *boost* our dataset and create more synthetic AD graph structures following similar patterns. These artificial datasets were used to train the model and evaluate it regarding the ability to generate graphs. Apart from evaluating the generation of graphs, we evaluate the quality of the positions of the nodes, and we evaluated the quality of honeyusers. This last is evaluated by publishing on the Internet a game to play by attackers. This game has an AD structure enriched with honeyusers from our model and an AD structure with randomly placed honeyusers. The results of this game help understand if real attackers are more lured into the honeyusers placed by the deep learning model.

Results show that our DAG-RNN autoencoder model can generate new honeyuser-enriched AD graphs that are in average 80% similar to the original graph. It can also place honeyusers in *organic* positions 94% of the time. Preliminary results from the real-life game of attackers are inconclusive but suggest an attackers' tendency to prefer the DAG-RNN generated honeyusers.

The contributions of this paper are:

- A new DAG-RRN autoencoder model for extending Active Directory graphs with honeyusers.
- The first time that Bidirectional DAG-RNN models are applied to the domain of honeyusers generation.
- An evaluation of the model with a real-life experiment involving real attackers.
- A new public code implementation of the DAG-RNN model that only depends on Tensorflow 2.
- Public dataset of the AD graphs used for our model

The rest of the paper is organized as follows: Sect. 2 describes the related work; Sect. 3 describes the generation of the artificial dataset; Sect. 4 describes the details of the deep learning method; Sect. 5 describes the experiments to evaluate the model both in generation and real-life experiment; Sect. 6 shows and discuss the results; and Sect. 7 makes the conclusion.

2   Related Work

Active Directory (AD) has been analyzed as a target of advanced attacks due to its importance inside organizations and companies [6]. These attacks have motivated to detect attacks in AD using advanced machine learning methods. The most common approach is to analyze the system logs of AD to search for anomalies [35].

Among the solutions available to protect AD are hardening solutions and monitoring tools [17]. In this regard, the main tool for detecting malicious activities in an Active Directory is the Advanced Threat Analytics by Microsoft [38], which can detect abnormal activity in the domain and report results. There are tools to manage fake accounts, such as BlueHive [4], but it does not generate new honeyusers. A similar approach was taken by the tool DCEPT [5] that creates fake accounts in the memory of endpoint devices. Even though honeypots have been a known technology for years, to our knowledge, there is no research to automatically generate and insert honeypots in the structure of AD [47].

In other security domains, there have been attempts to use automation and machine learning methods to design honeypots. Techniques include the use of state machines to generate scripts for creating honeypots [30] for the open-source honeypot manager honeyd [40]. Reinforcement Learning has also been used for generating honeypot responses to extend the duration of the attacker's session. [13].

Honeypot placement has also been studied by Game Theory researchers modelling the interaction between attackers and defenders as a two-player game [45].

There are considerable previous works using Graph Neural Networks (GNN) for detection, generation, and classification of graphs [51]. One of the most prominent works in the area of generative graph networks is GraphRNN [52]. Authors generated the graph iteratively using two recurrent modules, one for nodes and the other for graphs. GraphRNN outperforms Graph Convolutional Networks on the generation of realistic looking undirected graphs.

Graph Variational Autoencoders were used to generate small undirected graphs in molecule modelling with good success [42]. The method, however, lacks good scaling and works with structures of predefined maximal size.

Graph Recurrent Attention Networks, as proposed in [31], showed great success in modeling protein data with results exceeding both GrapVAE and GraphRNN. The technique combined recurrent GNN with attention layers.

When taking the directionality of graphs into account to create Directed Acyclic Graphs (DAG), several works proposed models that obtained good performance. In the area of logical formulas, custom RNN cells were used to analyze a DAG structure of logical formulas and their simplification as shown in [25]. Another work used a DAG to DAG framework learnt the satisfiability of formulas in propositional logic (SAT) [1]. Both works used the Encoder-Decoder architecture built on top of the graph recurrent cells.

There are several libraries implementing models of graph neural networks. Deep-GraphLibrary [49] is used for efficient graph storing and preprocessing. It is well integrated with NetworkX [19]. The Spektral library [16] is a Tensorflow library implementing mostly Graph Convolutional and Polling layers. Another library introduced by DeepMind is [3], which is based on Sonnet framework.

To the best of our knowledge, there are no publications using generative models for honeypot generation. The closest work is NeuralPot [43] that used Generative Adversarial Networks and AutoEncoders to generate the network traffic of the industrial Modbus honeypot. There are no research on generative models for honeypot generation, as it is proposed in this paper.

3 Dataset

Production environments of AD usually have sensitive Personally Identifiable Information (PII) from users that may be even regulated in the EU by the General Data Protection Regulation (GDPR) [29]. For this reason, it is very hard and maybe legally impossible to obtain good datasets of real AD structures from organizations. This posed a significant restriction on our gathering of data points for training and evaluating.

We solved the issue by obtaining a few samples of real AD structures by signing Non-disclosure Agreements (NDAs) with some organizations and then using these samples for *boosting* the generation of artificial datasets. The artificial datasets are generated by maintaining the same general characteristics of the real AD, aided in the selection of features by security experts.

Based on a few real AD samples obtained, we created four artificial datasets of graph structures which only differ in the number of nodes and edges. Each dataset contains a large number of artificial graphs with the similar number of nodes. All graphs are valid Directed Acyclic Graphs that follow the learned restrictions found in the real AD structures, such as which groups have more users.

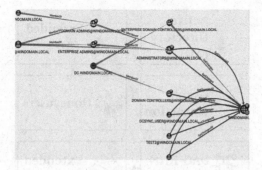

Fig. 1. Example of Active Directory (AD) Data extracted with the Sharphound tool and visualized using the Bloodhound tool. The extracted data is used to create a graph of the AD as input for the graph generation framework.

3.1   Extracting Active Directory Data

The existing AD structure has to be extracted before using the proposed framework for extending it with honeyusers. External tools such as Sharphound [48] can be used for the task. The example output of Sharphound in Fig. 1 shows the graphical structure of the nodes and relations in the AD. Unlike Sharphound, in this work we do not use multiple edge types, which further relaxes the problem and simplifies the graph.

We processed the real ADs by filtering them, so we only retain five node types. These node types are the only ones used in the artificial datasets: *User*, *Computer*, *Domain*, *OrganizationalUnit* (OU), and *Group* (in this context equivalent to Security-Group). For each real AD, we extracted these types of nodes and their relationships to create subgraphs.

To generate each of the four artificial datasets from these subgraphs, we used the random DAG generation of the NetworkX library [19]. All generated graphs in each dataset have the same node-to-edge ratio as the real AD structures they are based from, and they also have the same node types. Table 1 shows the properties of the datasets. The main difference between them is the size of the graphs.

Table 1. Comparison of artificial datasets in terms of number of graphs, number of nodes, mean amount of vertices and mean amount of edges. The number of edges for individual graphs is sampled from a Gaussian distribution using parameters estimated from real AD structures [33].

Dataset	graph size	# samples	Mean $|V|$	Mean $|E|$
AD15	15	2,000	12.51	19.02
AD50	50	2,000	39.88	65.49
AD150	150	2,500	115.11	192.49
AD500	500	1,000	353.36	600.17

We assume that, for a certain user, the number of relationships to other nodes is an important criterion that influences why an attacker chooses to attack that user. Therefore, the final usefulness of a honeyuser node for being a good target is related to how many connections it has and to which nodes.

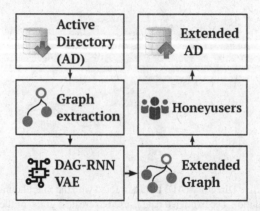

Fig. 2. Diagram of steps in the Graph Generation framework. First, it extracts a graph structure from an existing Active Directory (AD). Second, it generates an embedding of the features. Third, it processes the graph with a DAG-RNN Variational Autoencoder. Fourth, it predicts the locations of honeyusers in an extended graph. Fith, it enriches the honeyusers with additional attributes. Sixth, it inserts the honeyusers in the original AD [33].

Dataset Separation. Each one of these datasets of artificial graphs was independently splitted in 4/5 parts for training and validation, and 1/5 for final testing. The testing part was never used until the final evaluation. The training/validation sets were shuffled every time there was a new training interaction.

4   Graph Generation Framework

The graph generation framework is our proposal to generate nodes (honeyusers) that have connections to other nodes in the original AD graph in a topologically meaningful way.

The framework (i) processes AD graphs into a suitable representation by (ii) embedding the features of the nodes; (iii) uses a bidirectional sequential RNN encoder to generate a latent space; (iv) uses a decoder to sample nodes from the latent space and generate the estimated adjacency matrix of the new nodes; (v) enriches the features of the nodes to be inserted in the original AD structure; and (vi) extends the original AD with new users. Figure 2 shows a diagram of the proposed framework.

4.1   From AD to Input Matrices and Embeddings

The first step of the framework is to convert the AD graph into a representation that is good for processing. First, the AD graph is filtered to create a subgraph with only six types of nodes, a process described in Sect. 3. This smaller subgraph is then processed to create a topologically aware vector of nodes using a variation Depth-first search (DFS) [34] implemented by NetworkX [19].

This subgraph of the AD, now sorted, is used during the training time to generate datasets with similar topology, as it was described in Sect. 3.

From all artificial graphs in the mini-batch, we create two matrices X and A. The shape of X, called adjacency matrix, is (m, n, f) and the shape of A is (m, n, n), with m being the size of the mini-batch, n the size of the graphs after padding, and f is the size of the feature vector (6 in our case as there are 6 node types). The content of matrix X is, for each node of each graph, a one-hot encoding of the node type. Matrix A represents, for each graph, when two nodes are connected, and it is a lower triangular binary matrix, with $A \in \{0, 1\}^{n \times n}$.

The matrix X is used as input to an embedding layer. The embedding gives a space where the node types are better related and can be used by the RNN model. The embedding layer receives as input the matrix X and learns an embedding for the node features. The output of the embedding layer is the matrix X' with the same shape as X.

During the training time, the matrices X and A are padded to have the same number of nodes so they can be processed. The padding nodes are masked during the whole training.

4.2   The DAG-RNN Variational AutoEncoder

After the original AD structure has been converted to matrices X' and A, they are processed to obtain a latent space representation of nodes that follow the characteristics of the original AD graph. The goal is to train an autoencoder [28] that learns this representation and is able to generate new graphs with similar types of nodes and similar relationships between the nodes. For this, we created a DAG-aware RNN Variational Autoencoder model.

The general structure of the DAG-RNN [1] Variational Autoencoder, as shown in Fig. 3, uses as input the matrix X', A and a new topologically reversed matrix A^T, that is used later for the bidirectional part of the DAG-RNN.

The DAG-RNN outputs a final matrix H and internally works with two hidden state matrices. Matrix $\overrightarrow{H}$ that are the hidden states in the *forward* direction, and matrix $\overleftarrow{H}$ that are the hidden states used in the *reverse* direction. Figure 4 shows a diagram of the folded recurrent cell.

The output matrix H is used as input to two Multi-Layered Perceptron networks that learn the parameters μ and σ of a probability distribution representing the latent space of the nodes.

The decoder part of the autoencoder has two inputs. The first input is the vector z, sampled from the probability distribution (the required number of nodes to sample is predefined). The second input of the decoder is the matrix H which is an embedding representation of the nodes in the original graph.

The sampled representations of new nodes z are then paired with a Cartesian product to the representations in H. Each couple of nodes is given as an input to a MLP that is trained to decide if the proposed relation between two nodes should be kept as an edge or not.

With respect to the previous hidden states, the DAG-RNN works differently than the traditional RNN, where the previous states refer to the previous states in the sequence that is inputted. The DAG-RNN has one or more topological orders for the sequence and therefore there is a new meaning to the concept of *previous state*.

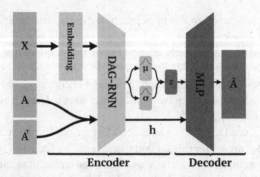

Fig. 3. Overview of DAG-RNN Variational Auto-Encoder model. There are two inputs: matrix X with on-hot encoded classes of graph nodes and adjacency matrix A. Rows in both matrices are ordered based on a topological ordering of nodes in G. The model consists of an Encoder (Embedding layer and DAG-RNN layer) which projects the inputs in latent space representation z and a Decoder (MLP) which reconstructs the adjacency matrix $\hat{A}$ [33].

DAG-RNN Encoder. In the core of the DAG-RNN layer are two Gated Recurrent Units [7], each processing edges in one direction. Figure 4 shows the folded recurrent cell and Fig. 5 shows the unfolded version. Each unit receives an input X' and a recurrent connection with a previous state $\overrightarrow{H}$ (the matrix used in the reverse direction is called $\overleftarrow{H}$). Contrary to traditional RNN cells where there is only one previous state that is taken from the previous unit, in a DAG-RNN, there are two main differences. First, the recurrent hidden state is not taken from the output of the previous unit in the sequence but from the output of units that correspond to *topologically-sorted* previous nodes in the graph structure (parents).

The second difference is that there can be an arbitrary number of previous states given that the nodes can have multiple predecessors in the graph. Therefore, the previous state of each DAG-RNN unit is the aggregated sum of all hidden states of all topologically sorted nodes that are *direct predecessors* of the current node in the graph.

The aggregation of hidden states is done by a sum of the Haddamard product of the matrices A and $\overrightarrow{H}$ [22]. The Haddamard product is an element-wise multiplication. For the reverse direction in the RNN, the matrix A is first topologically reversed (transposed) into a new matrix A^T which is used in a multiplication with matrix $\overleftarrow{H}$. The matrices A and A^T act as a mask selecting only those nodes that are topological predecessors of the current node. Equation (1) shows how each previous hidden state is computed for the forward direction of the RNN. Equation (2) shows the reverse direction case.

$$\overrightarrow{h_{t-1}} = \sum A \circ \overrightarrow{H} \tag{1}$$

$$\overleftarrow{h_{t+1}} = \sum A^T \circ \overleftarrow{H} \tag{2}$$

Let $G = \langle \mathcal{V}_G, \mathcal{E}_G \rangle$ be each of the graphs used as input for our method, with $\mathcal{V}_G$ being the set of vertices of G and $\mathcal{E}_G$ being the set of edges. We can say that by

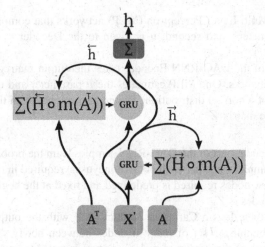

Fig. 4. Bi-directional DAG-RNN layer. Core of the layer is an off-the-shelf GRU cell which is fed with the previous states derived from the graph topology. A reversed graph for the bi-directional processing is also represented, using the matrix A^T instead of A to generate the hidden states. The outputs of both directions are summed [33].

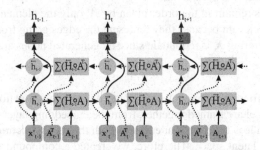

Fig. 5. Bi-directional DAG-RNN layer unfolded in time. For each timestep, the previous states are aggregated using the topology of the input graph. Sum of both directions is produced as output in each timestep.

topologically ordering the nodes of G it is guaranteed that $\forall v_i \subseteq V_G$, all of its predecessors have been already been processed in timestamps $t < i$.

The aggregation of hidden states of topologically sorted nodes gives the DAG-RNN the characteristic of generating new nodes that will be connected to other nodes in a *similar way* as nodes in the original graph. This is an important feature of the DAG-RNN that makes it possible to generate honeyusers that will be part of the most populated groups in the AD.

Our DAG-RNN implementation is bidirectional to capture the complete structure of the graph, given that processing a graph in only one direction may loss some information on successor nodes.

Each of the RNN cells produces one output for each of nodes in the sequence. In our implementation, the two outputs are summed.

The output of the DAG-RNN is the matrix H of node embeddings (hidden states) for all nodes of all graphs in the mini-batch. This output is used in two ways. First, it is given to both Multi Layer Perceptron (MLP) networks that compute the probability distribution parameters, and second, it is given to the Decoder to compute the final adjacency matrix $\hat{A}$.

The last part of the DAG-RNN Encoder takes the output matrix H and trains two standard MLP networks. One MLP estimates the μ parameter and the other estimates the σ parameter of a normal distribution. The nodes sampled from this distribution are represented by the matrix Z.

Decoder. The decoder part of the DAG-RNN samples from the probability distribution of the nodes. It samples once per each node (honeyuser) required in the final AD graph. The number of new nodes required is predefined and fixed at the beginning. The matrix of all requested nodes is Z.

The decoder then does a Cartesian product of Z with the output of the encoder H. Effectively obtaining a set of pairs of nodes between newly sampled nodes and original nodes. Each of these pairs of nodes is in turn inputted to an MLP that is trained to estimate if the pair should be kept or not. The output adjacency matrix $\hat{A}$ has the estimations if the nodes should be connected to each of the original ones. The MLP has as output a sigmoid activation, and a threshold value was trained to find the best results. The trained threshold was around 0.2.

Since the nodes remain in the order given by A, only the elements below the main diagonal of $\hat{A}$ are relevant because they represent the edges going from previous nodes. Note also that generating $\hat{X}$ is redundant since all generated nodes are of the same type: User.

Loss Functions. To train the model, we need to define a loss function. However, since all parts of the model are trained together, from the embedding layer to the MLP of the decoder, we need a loss function that captures both the reconstruction of the graph and the creation of the latent space. Therefore, we created a compound weighted loss that is the sum of two functions: a reconstruction loss and a latent loss.

Reconstruction Loss. The reconstruction loss estimates the information loss during the autoencoding process using the Focal Loss [32], which was originally designed for image classification problems. Focal Loss is a modification of the binary cross-entropy loss for problems with highly imbalanced classes. In our model, the presence or absence of an edge in $\hat{A}$ is treated as a binary classification.

In classic binary cross-entropy, the misclassification of either class is treated equally. This is well suited for domains where both classes are balanced, but it is not suited for our case since the the decision problem of pairing new nodes to AD nodes is highly imbalanced towards the class "No edge".

To mitigate this problem, we use the Sigmoid Focal loss (FL), shown in Equation (3).

$$FL(p_t) = -\alpha_t(1 - p_t)^\gamma log(p_t) \tag{3}$$

FL is a modification of binary cross-entropy using the parameters α and γ to address not only the imbalance but also the difference in the difficulty of classifying both classes. "While the parameter α balances the importance of positive/negative examples, it does not differentiate between easy/hard examples" [32]. The γ parameter scales the classification difficulty of the minority class.

In FL, $(1 - p_t)^\gamma$ is a modulating factor, while pt is a notational convenience defined as:

$$p_t = \begin{cases} p & \text{if } y = 1 \\ (1-p) & \text{otherwise} \end{cases}$$

where y specifies the ground-truth class, p is the prediction, and $log(p_t)$ is a representation of the Cross-Binary Entropy, as described in [32].

Latent Loss. The second part of our loss function, known as *latent loss* estimates how the distribution of the latent representation differs from the Normal distribution. For this, we used the Kullback-Leibler Divergence [24], shown in Eq. 4. D_{KL} measures the distance between the latent distribution Q and a Normal distribution as the prior P.

$$D_{KL}\left(P\|Q\right) = \sum_{x \in X} P(x)log(\frac{P(x)}{Q(x)}) \tag{4}$$

where X is the probability space, and $P = N(0,1)$.

The final combined loss function is defined as a weighted sum of the Focal loss and the latent loss as shown in Eq. 5.

$$\mathcal{L} = \frac{n^2 FL\left(A, \hat{A}\right)}{2} + |z| D_{KL}\left(N\left(z_\mu, z_\sigma^2\right) \| N\left(0,1\right)\right) \tag{5}$$

where n is the number of nodes, A is the original adjacency matrix, $\hat{A}$ is the estimated adjacency matrix, and z_μ and z_σ are the estimated parameters of the normal distribution. The Focal Loss is divided by two since we only estimate half of $\hat{A}$, that is a lower triangular matrix.

4.3 Honeyuser Attributes Generation

Even though the new graph with the positions of honeyusers was already generated by the DAG-RNN Autoencoder, the values of their attributes in the AD structure are still empty, such as name or phone. To be realistic, these attributes need to be completed.

There are two types of attributes: independent of the position and dependent of the position. Dependent of the positions are, for example, the *Distinguished Name* (DN). The DN is built by concatenating all the *Relative Distinguished Names* (RDN) from the node to the root node [12]. A sanity check is done to guarantee that some properties of AD are not violated, such as a honeyuser belonging to two Organizational Units. In case a node placement occurs which violates the constrains of the AD, it is discarded to ensure the consistency.

The independent attributes can be randomly generated using external tools. For this, we used the tools Faker [14] and FakeNameGenerator [10]. During this process, we checked that the generated attributes did not reveal the presence of the honeyuser, since some attributes such as *last logon*, should be filled by the AD server. The most important addition to the new nodes are the edges to the other objects in the AD as estimated by the DAG-RRN Autoencoder. This addition should be done both in the user's nodes and the group's nodes.

Insertion in the Original AD Structure. Once both the placement and attributes of the honeyuser were generated, they should be inserted in the AD server and registered as honeytokens to any potential monitoring service. The insertion is done using Powershell cmdlets or LDAP addition queries.

4.4 Implementation and Complexity

To the best of our knowledge, there was no implementation of a Python based Tensorflow 2 compatible DAG-RNN framework. The Graph Nets library [3] contains tools which simplify the building of such models but requires using the in-house library Sonnet. Our implementation is based purely on Tensorflow 2 and Keras frameworks and it is compatible with standard Tensorflow pipelines. Since it uses the Tensorflow back end for all computations, it can be used on CUDA GPUs.

Regarding the scalability of the implementation, it uses the memory sub-optimally. In particular, the implementation works with the adjacency matrix A represented as a dense matrix, but due to the ordering $\mathcal{O}_\mathcal{G}$, the matrix is lower triangular. One possible future improvement is to use the concept of Sparse Tensors introduced in Tensorflow 2, which allows for a more efficient representation.

Sequential processing of the nodes based on the topological ordering results in time complexity $\mathcal{O}(N)$. Node v can be processed only after all its predecessors have been already processed by the model. Each pair of nodes (u, v) is processed in the decoder which means the memory complexity is quadratic in the size of the input.

5 Experiments

The model was trained to find the best hyperparameters and it was evaluated in three different ways. First, the model was evaluated on its ability to encode and reconstruct existing graph structures precisely, without adding new nodes. Second, the model was evaluated on its ability to extend graphs using the DAG-RNN VAE. Third, the model was evaluated on its capacity to generate honeyusers that attract attackers in real-life AD systems.

5.1 Experimental Setup

The hyperparameters of the model were trained with a mixture of grid search and heuristic expert knowledge based on previous works.

The embedding layer output dimension is 6, since there is no need for dimensionality reduction. The GRU cell in the encoder consists of 64 units, while the two MLPs that estimate μ and σ have 32 hidden units each.

The MLP encoder has 3 hidden layers with 64, 64 and 32 units respectively, and it uses ReLU activations. The single output unit uses sigmoid activation.

The whole model, from the embedding layer, DAG-RNN encoder, to the MLP decoder is trained simultaneously by minimizing the compound loss function.

In all cases, the Adam [27] optimizer is used for the training with an exponentially decayed learning rate for fine tuning of the parameters towards the end of the training period.

Initial weights of the DAG-RNN and Decoder MLP are obtained using the Glorot uniform initializer [15] except for the hidden dense layer for estimating σ where initial weights are set to 0.

The model is trained using a single computer with 32 GB of RAM and Nvidia Titan V GPU card with 12 GB of memory. All code is free software and accessible for the community [1].

5.2 Graph Reconstruction

As part of the training process, the evaluation of the graph reconstruction was used as one of the measures to determine if the model performed correctly. The graph reconstruction measures the complete generation power of the model to create new graphs that are similar to the original AD graph. This is done by comparing the input embedding adjacency matrix A with the reconstructed counterpart $\hat{A}$.

The two matrices were compared element-wise. Therefore, the confusion matrix was built in the following way:

Table 2. Composition of the confusion matrix for graph reconstruction metrics.

	$\hat{A}_{i,j} = 1$	$\hat{A}_{i,j} = 0$
$A_{i,j} = 1$	TP	FN
$A_{i,j} = 0$	FP	TN

From this confusion matrix, we computed the metrics Precision, Recall, F1 score, and the area under the Precision-Recall Curve (PR AUC). Due to the imbalance in this decision problem, as discussed in Subsect. 4.2, we can not use accuracy as a metric [44]. The comparison is done for nodes of the same type, and for the generated nodes the type is copied from the original nodes.

5.3 New Nodes Generation

The second evaluation used during the training was for the generation of new nodes, which answered the question: are the nodes connected in a meaningful way? We implemented two metrics: Edge Validity Ratio (EVR), and Mean Edge Count Ratio (MECR).

These metrics were chosen because in generative models, there is no ground truth to compare with, so our metrics are inspired by common metrics of GANs [18].

---

[1]https://github.com/stratosphereips/AD-Honeypot.

Edge Validity Ratio. EVR is the ratio between the amount of valid edges generated for a node and the total amount of generated edges. A valid edge is one that is possible for the constrains of an AD system, such as that a node can not be connected to two Organizational Units. Equation 6 shows the computation of the EVR for a node. Where $\delta^-(v)$ is the amount incoming edges of a node v and $\delta^-_{valid}(v)$ is the amount of *valid* incoming edges for the same node.

$$EVR(v) = \frac{\delta^-_{valid}(v)}{\delta^-(v)} \tag{6}$$

With the EVR of every node, we can compute the EVR of the graph. Equation 7 shows this computation, that is an average using $|V|$ as the number of nodes.

$$EVR(G) = \frac{1}{|V|} \sum_{v \in \mathcal{V}} EVR(v) \tag{7}$$

Mean Edge Count Ratio. The Mean Edge Count Ratio measures the mean number of incoming edges for a user node type (in particular for node type User). Given the mean in-degree of user nodes in the input graph:

$$\delta^-_I = \frac{1}{|V_{User}|} \sum_{n \in V_{User}} \delta^-(n) \tag{8}$$

and mean in-degrees of generated nodes (all of them are of type User):

$$\delta^-_G = \frac{1}{n} \sum_{i=1}^{n} \delta^-(v_i) \tag{9}$$

Then the MECR is a similarity metric defined as:

$$MECR(I, G) = \frac{\min\left(\delta^-_G, \delta^-_I\right)}{\max\left(\delta^-_G, \delta^-_I\right)} \tag{10}$$

MECR metric computes the similarity between two graphs based on the mean in-degree of the nodes of the same type. The best value for this metric is 1, where the user nodes in the extended graph have in average the same number of incoming edges as the original. Only which are connected to the rest of the graph (they have at least one edge) are included in this metric.

5.4   Evaluation of Nodes as Honeyusers

The final goal of the research was to generate honeyusers, which means fake users that attackers chose in an attack on an Active Directory system. To measure how valid are the nodes for a security defense, we executed a real-life attacking game.

By definition, any interaction with a honeyuser is considered an attack, since there is no official purpose to interact with the user [41]. Therefore, anytime an attacker selects a honeyuser, a detection is registered. On the contrary, if the attacker does not choose any honeyuser, the attack was not detected.

For this experiment, we set up two real Windows Active Directory systems on the Internet. One of them has our honeyusers placed in strategic positions. The other has the same honeyusers (number and features) but placed in random positions. The AD resembles an organization with 100 users. This way, we could compare if attackers selected our honeyusers more in comparison with the random positioned honeyusers.

The second AD positioned the nodes randomly. The position was determined at random by choosing if two nodes should be connected by following a uniform distribution. However, the number of edges of the nodes was selected from the probability distribution of the mean number of edges in the original AD graph.

Real-life Experiment Protocol. The protocol to play the game was as follows. First, users were contacted by social networks to invite them, focusing on the security industry. Second, the users were directed to a webpage where the game was explained [redacted for reviewers]. Third, the users played the game by connecting to our AD servers on the Internet. The three answers to the questions of the game were answered on the webpage.

Our game has two features from behavioral economic science. First, the participants are given a negative reward. They start the game with 3 USD, and every time they select a honeyuser for an answer, 1 USD is discounted. This motivates them not to lose money. Second, we donate the money that remains to a charity. The donation to a charity further increases the participants' motivation since it compensates the probable discomfort of attacking a server [46]. This complete approach proved to be more motivating than positive reinforcement via rewards and makes the participant more cautious [23].

Finally and more importantly, the selection of which AD server the attacker uses is randomly done on the webpage, controlling that there is no bias in the assignment of the AD servers.

During the evaluation, a task is solved successfully if the selected target is a legitimate domain account, if it is not a honeypot, and if it fulfills the given task. This means that if the task at hand was, for example, to access data from the Finance Unit of the organization, the selected user should have permission to do that in the AD.

A task is not solved successfully if the selected user is a honeyuser or a legitimate domain account that does not fulfill the given task. For each unsolved task, 1 USD is taken from the attacker's credit. Any remaining credit after the end of the game is donated to the charity. The tasks were picked before generating any honeypot.

6   Results and Discussion

Three different types of evaluations were done, which results are reported in this section. First, the model was evaluated for its ability to generate graphs with the DAG-RNN VAE similar to the original AD. Second, the model was evaluated for its capacity to place honeyusers in places that are similar to the current users of the AD. Third, the model was evaluated for its capacity to attract real attackers by publishing an online game for attackers.

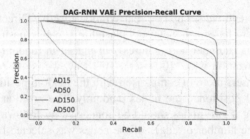

Fig. 6. Comparison of Precision-recall Curve on testing data for the reconstruction of the original graph by the DAG-RNN. Graphs up to 150 nodes have enough reconstruction precision to be useful, but graph of 500 nodes are not reconstructed precisely enough [33].

6.1 Results of Graph Reconstruction

For graph reconstruction, there are four sets of results for each of the datasets: AD15, AD50, AD150 and AD500. Table 3 shows that the model achieves nearly 80% precision in the testing set of AD15, AD50, and AD150. However, it only reaches 51% precision for large datasets of 500 nodes, suggesting that in larger graphs the ability to reconstruct the graph degrades. The F1-score measure reaches 84% for middle-size graphs and is close to 60% for large graphs. Figure 6 shows a comparison of precision-recall curves, where it is seen that the largest dataset AD500 has a drop in performance.

These results suggest that our method can reconstruct graphs with enough precision up to 150 nodes and struggles with graphs of 500 nodes. These results are of enough quality to be useful in the generation of new users.

Table 3. Evaluation of the capability for graph reconstruction of the DAG-RNN VAE model [33].

dataset	precision	recall	F1-score
AD15	80.93%	94.5%6	87.22%
AD50	79.94%	89.48%	84.44%
AD150	80.38%	45.53%	58.13%
AD500	51.85%	72.6% 7	60.52%

6.2 Results of New Nodes Generation

The results of the evaluation of the generation of new nodes are also separated for each of the four datasets. Table 4 shows the EVR and MECR metrics applied to all datasets. These metrics measure how *organic* was the addition of the new honeyuser. The Edge Validity Ratio (EVR) was best for graphs of 50 nodes, with a precision of 80% and and F1-score of 84%. Graphs of 150 and 500 nodes had an F1-score close to 60% meaning that for larger graphs we are not generating edges that are in the same amount or connected to the same groups as the other users in the AD.

However, with close to 84% F1-score, we can expect to generate new nodes for graphs of middle size that are *organic* enough to be very similar to the other users of the original graph.

Table 4. Similarity metrics for the graphs generated by the DAG-RNN VAE per dataset [33].

Dataset	EVR	MECR
AD15	68.38%	77.53%
AD50	72.21%	95.42%
AD150	69.18%	92.42%
AD500	58.86%	95.23%

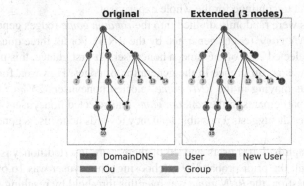

Fig. 7. Example of a graph extended with DAG-RNN VAE (right) from an original graph (left) from dataset AD15. There are three user nodes with ids 15, 16, and 17 (in red) inserted in the left graph. (Color figure online)

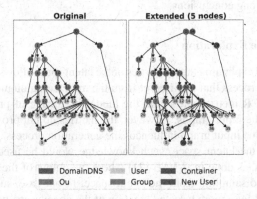

Fig. 8. Example of a graph extended with DAG-RNN VAE (right) from an original graph (left) from dataset AD50. There are five user nodes with ids 50–54 (in red) inserted in the left graph. Node 54 is disconnected from the graph and is to be discarded as described in Subsect. 4.3 [33]. (Color figure online)

Previous work on DAG generation worked on graphs up to 10 nodes [1], while our approach was able to work on graphs up to 150 nodes with fair results. This was possible because our specific task of honeyuser generation needed less precision to work.

Figure 7 shows an example of our generated graph after inserting the new honeyuser node. It can be seen how both graphs are fairly similar and that there are three new users (in red) added to the graph on the right. Figure 8 is an example of a generated graph of 50 nodes with honeyusers inserted.

6.3   Results of Evaluating Nodes as Honeyusers

The evaluation of the value of the nodes as real honeyusers has been hard to estimate. This is because it is hard to find real attackers willing to play our game. With ten participants in the study so far, the results are not statistically significant, but they show trends that we expect to continue for the whole experiment.

The participants were randomly divided into the *random game* (edges generated by random) and the *RNN game* (edges generated by the RNN). For all three questions, in total, both groups selected 25% of the time, a honeyuser. At first glance, this may seem to suggest that there is no difference in the generation of edges. However, for the first question participants playing in the *RNN game* selected honeyusers 25% of the time, compared with the participants in the *random game* that selected honeyusers 12.5% of the time. This last result suggests a possible tendency towards honeyusers generated by the RNN.

The original graph had 100 users and 20 of them were inserted honeyusers. This means that by default the prior probability of choosing a honeyuser was 16.6%. However, for the first question, the *RNN game* was over this threshold by reaching a 25% of honeyusers hits, suggesting that honeyusers generated by the RNN maybe be selected more than expected. At the contrary, participants in the *random game* were below this threshold with 12.5% of honeypot hits, which shows that the amount of data is not enough to draw strong conclusions.

6.4   Latent Space Exploration

Understanding and explaining the properties of the latent space allows a better evaluation of the generation process that is different from the more common metrics shown in 4.

EVR and MECR metrics show that the model is capable of producing meaningful extensions of the input graphs. However, the metrics do not provide insights about which features of the input graph influences the generation process. If it would be possible to influence the latent space, such knowledge would be used for tweaking the generation process, as done in β-VAE [21] where the output of the Decoder was controlled by informed sampling from the latent space. In our case, such tweaking could potentially be used for a more targeted creation of the honeyusers, possibly resulting in an even higher detection rate.

As a first step, we explored the latent space searching for an explanation of how the nodes were formed. Figure 9 shows the latent space representation of the complete AD150 dataset graph nodes. The latent space, originally having a dimensionality $|z| = 32$, is being transformed in the 2 dimensions with he UMAP algorithm [36]. UMAP attempts to preserve the local structure of the data during the dimensionality reductions which naturally groups samples in clusters.

It is clear that our Encoder is able to project the graph nodes in the latent space so that the node clusters are easily separable. Such results could be directly used to

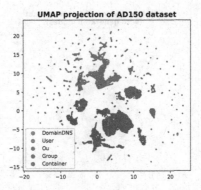

Fig. 9. UMAP projection of the latent space representation of nodes from graphs in the AD150 dataset into two dimensions. Clusters following the node types are clearly separable.

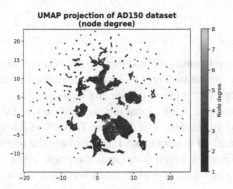

Fig. 10. UMAP projection of the latent space representation of samples from AD150 with node-degree highlighted.

influence the generation process with bias in the sampling process. However, it is not sufficient to determine the relation between the node properties and its latent space projection, it is necessary to understand which features played a more important role in the projection. Given the fact that only the *node type* is used as a node feature vector $x \in X$, we can presume that the node position in the input graph plays a significant role in the latent space representation.

One might be lead to conclusion that the clusters are naturally following the node-degree representation, since it is such an important property. However, when we highlighted the node-degree in the UMAP graph, now shown in Fig. 10, it can be seen that the nodes with same node-degree are equally scattered so we can conclude the Encoder captured more complex structural information in the graphs.

To explain which other structural properties of the nodes were used in the latent space (apart from node-degree) we can go *back* from the generated honeyusers, and project their latent vectors into the latent space. Figure 11 shows this projection, where it can be seen that every type of node is included in the latent space. Such an explanation opens the door for a future improvement where a constrain in the latent space may be used for forcing the variational autoencoder to generate nodes of only one type, such as type *user*.

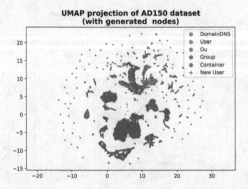

Fig. 11. UMAP projection of latent space representation of node from graphs in the AD150 dataset into 2 dimensions.

7 Conclusion

This work shows a deep learning method that ingests real Active Directory (AD) structures and generates a similar structure by inserting honeyusers (fake users) that can be used to detect attackers. Our method specifically chooses where to place honeyusers in the AD structure by using a bidirectional topologically sorted DAG-RNN autoencoder that was implemented for the first time as a pure Tensorflow 2 library.

The core of the method is a framework that embeds the topological node features, trains a DAG-RNN Encoder which transforms the graph into a latent space, and uses a Decoder to generate new graphs with inserted honeyusers in *organic* positions in the original AD graph.

The model was evaluated in three ways. First, in its ability to generate similar graphs, where it shows that in graphs up to 150 nodes it can get 80% precision. Second, in its ability to place the nodes organically, where it obtained a Mean Edge Count Ratio of 92%. Third, in its capability of generating honeyusers that are attractive to attackers by creating a public game for real attackers. The results of the game are still inconclusive given the small number of participants, but preliminary results seem to suggest that the nodes placed by the RNN are selected slightly more.

Our Future Work: will be to finish the experiments with real attackers, to estimate the node type from the embedding directly, to include more features from the nodes in the first embedding layer, and to include a feature related to the attractiveness of the original groups for the end-user of the framework to bias the placement towards those groups.

Acknowledgements. We acknowledge the support of NVIDIA Corporation with the donation of a Titan V GPU for this research. We would also like to thank the Stratosphere team for their support.

References

1. Amizadeh, S., Matusevych, S., Weimer, M.: Learning to solve circuit-SAT: An unsupervised differentiable approach. In: International Conference on Learning Representations (2019). https://openreview.net/forum?id=BJxgz2R9t7
2. de Barros, A.P.: Res: Protocol anomaly detection ids - honeypots (2003). https://seclists.org/focus-ids/2003/Feb/95
3. Battaglia, P.W., et al.: Relational inductive biases, deep learning, and graph networks (2018)
4. Berg, L.: BlueHive (2019). https://github.com/leeberg/BlueHive
5. Bettke, J., Stewart, J.: DCEPT: An Open-Source Honeytoken Tripwire (2016). https://www.secureworks.com/blog/dcept
6. Case, D.U.: Analysis of the cyber attack on the ukrainian power grid. Electricity Information Sharing and Analysis Center (E-ISAC) p. 388 (2016)
7. Cho, K., et al.: Learning phrase representations using rnn encoder-decoder for statistical machine translation (2014)
8. Cimpanu, C.: Fortune 500 company ntt discloses security breach (2020). https://www.zdnet.com/article/fortune-500-company-ntt-discloses-security-breach
9. Cimpanu, C.: Hackers breached a1 telekom, austria's largest isp (2020). https://www.zdnet.com/article/hackers-breached-a1-telekom-austrias-largest-isp
10. CorbanWorks: Fake name generator (2006). https://www.fakenamegenerator.com
11. Crabtree, J.: Active directory attacks hit the mainstream (2020). https://www.darkreading.com/endpoint/authentication/active-directory-attacks-hit-the-mainstream/a/d-id/1337405
12. Desmond, B., Richards, J., Allen, R., Lowe-Norris, A.G.: Active Directory: Designing, Deploying, and Running Active Directory. " O'Reilly Media, Inc." (2008)
13. Dowling, S., Schukat, M., Barrett, E.: Using reinforcement learning to conceal honeypot functionality. In: ECML/PKDD (2018)
14. Faraglia, D.: Faker (2012). https://pypi.org/project/Faker/
15. Glorot, X., Bengio, Y.: Understanding the difficulty of training deep feedforward neural networks. In: Proceedings of the International Conference on Artificial Intelligence and Statistics (AISTATS'10). Society for Artificial Intelligence and Statistics (2010)
16. Grattarola, D., Alippi, C.: Graph neural networks in tensorflow and keras with spektral (2020)
17. Grimes, R.A.: Honeypots for Windows. Apress (2006)
18. Guan, S., Loew, M.: Evaluation of generative adversarial network performance based on direct analysis of generated images. In: 2019 IEEE Applied Imagery Pattern Recognition Workshop (AIPR), pp. 1–5 (2019). https://doi.org/10.1109/AIPR47015.2019.9174595
19. Hagberg, A.A., Schult, D.A., Swart, P.J.: Exploring network structure, dynamics, and function using. In: Varoquaux, G., Vaught, T., Millman, J. (eds.) Proceedings of the 7th Python in Science Conference, pp. 11–15. Pasadena, CA USA (2008)
20. Han, X., Kheir, N., Balzarotti, D.: Deception techniques in computer security: A research perspective. ACM Comput. Surv. 51(4) (Jul 2018). https://doi.org/10.1145/3214305
21. Higgins, I., et al.: beta-vae: Learning basic visual concepts with a constrained variational framework. In: ICLR (2017)
22. Horn, R.A.: The hadamard product. In: Proc. Symp. Appl. Math. vol. 40, pp. 87–169 (1990)
23. Hossain, T., List, J.A.: The behavioralist visits the factory: Increasing productivity using simple framing manipulations. Manage. Sci. 58(12), 2151–2167 (2012). http://www.jstor.org/stable/23359584
24. Joyce, J.M.: Kullback-Leibler Divergence, pp. 720–722. Springer, Berlin Heidelberg, Berlin, Heidelberg (2011). https://doi.org/10.1007/978-3-642-04898-2_327

25. Kaluza, M., De Paolis, C., Amizadeh, S., Yu, R.: A neural framework for learning dag to dag translation. In: NeurIPS'2018 Workshop (2018)
26. Karlin, A.R., Bradley, M., Baldwin, M., Sagir, S.: What threats does ata look for? (2018). https://docs.microsoft.com/en-us/advanced-threat-analytics/ata-threats
27. Kingma, D.P., Ba, J.: Adam: A method for stochastic optimization (2014)
28. Kingma, D.P., Welling, M.: Auto-encoding variational bayes (2014)
29. Koch, R.: What is considered personal data under the EU GDPR? (2020). https://gdpr.eu/eu-gdpr-personal-data/
30. Leita, C., Mermoud, K., Dacier, M.: Scriptgen: an automated script generation tool for honeyd. In: 21st Annual Computer Security Applications Conference (ACSAC'05), pp. 12 pp.-214 (2005)
31. Liao, R., et al.: Efficient graph generation with graph recurrent attention networks. In: NeurIPS (2019)
32. Lin, T.Y., Goyal, P., Girshick, R., He, K., Dollár, P.: Focal loss for dense object detection (2017)
33. Lukas, O., Garcia, S.: Deep generative models to extend active directory graphs with honeypot users. In: Proceedings of the 2nd International Conference on Deep Learning Theory and Applications, DeLTA 2021, pp. 140–147 (2021)
34. Manber, U.: Introduction to Algorithms: A Creative Approach. Addison-Wesley Longman Publishing Co., Inc, USA (1989)
35. Matsuda, W., Fujimoto, M., Mitsunaga, T.: Detecting apt attacks against active directory using machine leaning. In: 2018 IEEE Conference on Application, Information and Network Security (AINS), pp. 60–65. IEEE (2018)
36. McInnes, L., Healy, J., Melville, J.: Umap: Uniform manifold approximation and projection for dimension reduction (2018). https://doi.org/10.48550/ARXIV.1802.03426, https://arxiv.org/abs/1802.03426
37. Metcalf, S.: Red vs. blue: Modern active directory attacks, detection & protection (2015). https://www.blackhat.com/docs/us-15/materials/us-15-Metcalf-Red-Vs-Blue-Modern-Active-Directory-Attacks-Detection-And-Protection-wp.pdf
38. Microsoft: Advanced Threat Analytics documentation (2015). https://docs.microsoft.com/en-us/advanced-threat-analytics/
39. Nurfauzi, R.: Active directory kill chain attack & defense (2020). https://github.com/infosecn1nja/AD-Attack-Defense
40. Provos, N.: Honeyd-a virtual honeypot daemon (2003). http://www.honeyd.org/
41. Provos, N., et al.: A virtual honeypot framework. In: USENIX Security Symposium. vol. 173, pp. 1–14 (2004)
42. Simonovsky, M., Komodakis, N.: Graphvae: Towards generation of small graphs using variational autoencoders (2018)
43. Siniosoglou, I., et al.: Neuralpot: an industrial honeypot implementation based on convolutional neural networks (4 2020). http://gala.gre.ac.uk/id/eprint/27976/
44. Thomas, C., Balakrishnan, N.: Improvement in minority attack detection with skewness in network traffic. In: Tolone, W.J., Ribarsky, W. (eds.) Data Mining, Intrusion Detection, Information Assurance, and Data Networks Security 2008. vol. 6973, pp. 226–237. International Society for Optics and Photonics, SPIE (2008). https://doi.org/10.1117/12.785623
45. Tian, W., et al.: Honeypot game-theoretical model for defending against apt attacks with limited resources in cyber-physical systems. ETRI J. **41**(5), 585–598 (2019)
46. Tonin, M., Vlassopoulos, M.: Corporate philanthropy and productivity: Evidence from an online real effort experiment. Manage. Sci. **61**(8), 1795–1811 (2015). https://doi.org/10.1287/mnsc.2014.1985

47. Valicek, M., Schramm, G., Pirker, M., Schrittwieser, S.: Creation and integration of remote high interaction honeypots. In: 2017 International Conference on Software Security and Assurance (ICSSA), pp. 50–55. IEEE (2017)
48. Vazarkar, R.: Sharphound (2016). https://github.com/BloodHoundAD/SharpHound3
49. Wang, M., et al.: Deep graph library: Towards efficient and scalable deep learning on graphs (2019)
50. Whittacker, Z.: Hackers went undetected in citrix's internal network for six months (2019). https://techcrunch.com/2019/04/30/citrix-internal-network-breach
51. Wu, Z., Pan, S., Chen, F., Long, G., Zhang, C., Yu, P.S.: A comprehensive survey on graph neural networks. IEEE Transactions on Neural Networks and Learning Systems p. 1–21 (2020). https://doi.org/10.1109/TNNLS.2020.2978386
52. You, J., Ying, R., Ren, X., Hamilton, W.L., Leskovec, J.: Graphrnn: Generating realistic graphs with deep auto-regressive models (2018)
53. Zetter, K.: Sony got hacked hard: What we know and don't know so far (2014). https://www.wired.com/2014/12/sony-hack-what-we-know

Crack Detection on Brick Walls by Convolutional Neural Networks Using the Methods of Sub-dataset Generation and Matching

Mehedi Hasan Talukder[1]([✉]), Shuhei Ota[2], Masato Takanokura[2], and Nobuaki Ishii[2]

[1] Department of Computer Science and Engineering, Faculty of Engineering, Mawlana Bhashani Science and Technology University, Tangail, Bangladesh
mehedi@mbstu.ac.bd

[2] Department of Industrial Engineering and Management, Faculty of Engineering, Kanagawa University, Yokohama, Japan
{ota,takanokura,n-ishii}@kanagawa-u.ac.jp

Abstract. The appearance of cracks is considered an initial sign of the deterioration of structures such as concrete and brick walls. Crack detection plays an important role in ensuring the safety and durability of structures. Conventionally, a maintenance engineer performs crack detection manually, which is laborious and time-consuming. Therefore, a systematic crack detection method is required. Among the existing crack detection methods, convolutional neural networks (CNNs) are more effective; however, CNNs often fail in the case of brick walls. There are several types of bricks, and some may appear to have cracks owing to their structure. Additionally, the joining points of bricks may appear as cracks; therefore, CNN fails. It is theorized that CNN performance can be improved if sub-datasets are generated based on the image attributes, and a proper sub-dataset is selected by matching the test image with the sub-datasets. In this study, sub-dataset generation and matching methods are proposed to improve the performance of crack detection in brick walls using CNN. CNN training is conducted with each sub-dataset generated by the proposed sub-dataset generation method, while crack detection is performed using a proper trained CNN that is selected using the proposed matching method. For numerical experiments, training datasets are first prepared by manual image cropping and rotation, after which the performance of crack detection is evaluated by cross-validation. Numerical experiments show that the proposed method improves crack detection in brick walls. This study will help to ensure the safety of structures as well as the safety of human life.

Keywords: Deep learning · Maintenance of brick walls · Color information

1 Introduction

Cracks are common defects of different types of structures, such as concrete walls, brick walls, pavements, and bridges. A crack is a break without the complete separation of the parts and is considered the initial sign of structural deterioration. Crack detection and

A. Fred et al. (Eds.): DeLTA 2020/DeLTA 2021, CCIS 1854, pp. 134–150, 2023.
https://doi.org/10.1007/978-3-031-37320-6_7

repair in the early stages have great significance in structural maintenance for ensuring the safety and durability of structures and preventing significant damage [1]. A large number of old structures, such as buildings, highways, and bridges, in developed countries, are increasing rapidly [1–3]. According to a report by the Road Maintenance Bureau of Japan, the number of old bridges in Japan is increasing [2]. There are 700,000 bridges, and 43% of these bridges will have been in service for over 50 years in 2023 [2]. According to a report by the American Society of Civil Engineers (ASEC), infrastructure is the backbone of the U.S. economy. However, the number of older structures is increasing rapidly. According to ASEC, 24% of school buildings are in poor condition [3]. In addition, 40% of the bridges in the USA are 50 years or older [3]. The average ages of these bridges are approaching the end of their design lives. One out of every five miles of paved roads is in poor condition [3]. Structures are becoming older, thus making crack occurrences more likely. Cracks may lead to significant damage or failure if proper actions are not taken to repair the cracks [4]. Therefore, regular maintenance is necessary to ensure the safety and durability of structures [5, 6].

Conventionally, a maintenance engineer is responsible for detecting and measuring cracks. However, there is a shortage of maintenance engineers. In Japan, approximately 50% of towns and 70% of villages have no technicians for maintenance [2]. For this reason, the crack detection and repair of structures using a systematic method has become an important research issue.

Various systematic methods are used for the purpose of crack detection. Widely used and popular crack detection methods include image processing- and vision-based methods [7–11], and deep learning methods, such as convolutional neural networks (CNNs) [5, 12–15]. Among the existing crack detection methods, it is known that current image processing methods for crack detection are only effective under certain environmental conditions [12, 13]. Currently, CNN-based methods are generally more effective for crack detection compared to other methods [5, 6, 16–19]. Therefore, CNN-based methods are popular and widely used for crack detection. However, CNNs fail under varied environmental conditions such as shadows, low brightness, and stains [5, 12, 20, 21]. Moreover, the performance of CNN methods is low for crack detection in brick walls [6]. This is because there are several types of brick walls with different colors. Additionally, the joining points of bricks may appear as cracks, potentially leading to false results at the time of crack detection [4]. Therefore, improving the performance of CNNs for crack detection in brick walls is an important research issue.

Crack detection using a CNN-based method involves the collection of a dataset of images used for training and is performed based on this dataset. A dataset containing images with and without cracks must be collected; these images of the training dataset must be labeled correctly into two classes: crack and no-crack classes. Then, the dataset can be used for CNN learning, and crack detection can be performed based on the learned CNN.

At the time of crack detection, a large dataset comprising images with different properties is required for CNN training. There are several types of bricks with different colors; therefore, crack detection in brick walls is difficult by CNN methods using a large training dataset. Some images of brick walls with different properties are presented in Fig. 1 [27]. In this figure, sample crack and no-crack images are shown in the first and second rows, respectively. These images of brick walls in Fig. 1 are collected from the different buildings of Kanagawa University, Japan.

From Fig. 1, in some cases, the joining points of bricks are dark and look like cracks, while in other cases, the bricks themselves look like cracks. For this reason, the performance of CNN methods is poor in detecting cracks in brick walls using one large dataset consisting of the images of different brick walls for learning.

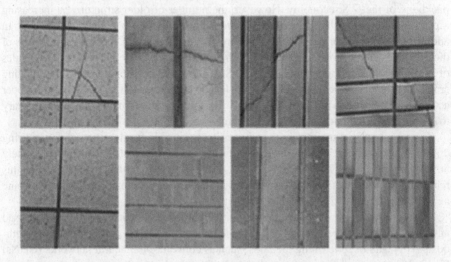

Fig. 1. Some images of different brick walls (Talukder et al. [27]).

Training a CNN using sub-datasets consisting of the images of similar bricks and selecting a proper sub-dataset for crack detection by matching the test image with the images of sub-datasets is theorized to improve the accuracy of crack detection in brick walls using CNNs.

Therefore, this study aimed to develop sub-dataset generation and matching methods to be used for CNN-based crack detection in brick walls. Sub-datasets generation and a proper sub-dataset selection by matching are conducted in the following ways to improve the accuracy of crack detection of brick walls:

1) Sub-dataset generation: Multiple sub-datasets are generated from one large dataset of images using the proposed sub-dataset generation method. Generated sub-datasets contain images of similar bricks. CNN learning is performed using these generated sub-datasets.

2) Sub-dataset selection by matching: After CNN learning, the matching method is used to select a proper learned CNN by matching the attributes of an input image with those of the sub-datasets. For crack detection, learned CNN is selected, which is learned by the sub-dataset comprising images of similar attributes to the test image.

Image attributes, such as color, brightness, and histogram, are used for image classification [22, 23]. RGB (red, green, blue) color provides strong indications of image classification and increases the image classification accuracy [22]. However, the images of brick walls contain sufficient information on red (R), green (G), and blue (B) values as shown in Fig. 1. Hence, the R, G, and B values of images are considered attributes for the sub-dataset generation and the matching methods.

The novelty and uniqueness of the proposed method lie in sub-dataset generation from one dataset and in the proper learned CNN selection by matching the color attributes of the test images with those of the sub-datasets. Sub-datasets were generated from a large dataset using the proposed sub-dataset generation method. Each sub-dataset contained images of similar color information, and was used for CNN learning. After learning, a proper sub-dataset was selected for crack detection using the proposed matching method. Therefore, the proposed method effectively improved the performance of the crack detection using the CNN-based method in brick walls.

2 Related Works

Many researchers have attempted to develop various techniques for crack detection. Widely used and popular crack detection techniques include deep learning methods, such as CNNs, and image-based methods, such as image processing-based, vision-based, filtering-based, and morphological operation-based methods.

Ozgenel et al. [6] used vision-based, image processing-based, and CNN-based methods for the purpose of crack detection. Their study showed that shadows and noises can affect the performance of crack detection using CNN methods. Cha et al. [5] performed crack detection using image processing and CNN methods; their study showed that the performance of crack detection using CNN methods is substantially affected by the noise created from different lighting sources and distortion. Huyan et al. [15] used a fast-region CNN-based crack deep network to detect sealed/unsealed cracks with complex road backgrounds. Their study states that crack detection performance is strongly influenced by image surface. According to this study, the performance of CNN is low for detecting cracks in images with certain complex background conditions, such as shadows and noise.

Zhang et al. [24] studied crack detection from images with complex textures while focusing on distinguishing between cracks and sealed cracks with identical widths and brightness. CrackNet was a CNN-based model presented by Zhang et al. [25] for crack detection where there were no pooling layers in the CrackNet architecture, unlike in the traditional CNN model. This architecture ensured accuracy up to the pixel level, as the image length and width remained unchanged in all layers. Yang et al. [26] used a CNN variation called a fully convolutional network (FCN) for the pixel-level segmentation of cracks from images of walls and pavements. The FCN model was trained using multiple types of crack images to predict pixel skeletons where the width of only one pixel was used to represent the crack segments. Talukder et al. [13] proposed a method consisting of a sub-dataset generation and matching-based CNN for crack detection in concrete walls, which used the brightness value of images for sub-dataset generation and matching as image properties.

Dais et al. [4] performed crack detection in different brick masonry using CNN and transfer learning. Their study showed that CNNs are more effective for crack detection compared to other methods. However, the performance is low for brick walls because there are several types of bricks, and the joining points of the brick walls are different in different walls, making crack detection more challenging.

Talukder et al. [27] proposed sub-dataset generation and matching methods for crack detection in brick walls using CNN. The authors showed that images of brick walls contain sufficient information on RGB color; hence, the R, G, and B values of images are used for the sub-dataset generation and matching methods. Talukder et al. [27] generated a large dataset of brick walls by using manual cropping containing 400 images from which sub-datasets were generated. However, dataset generation by applying only cropping is insufficient. Therefore, in this study, a large dataset is generated by applying image rotation with image cropping as image rotation is an effective method for covering different crack orientations [28, 29].

3 Proposed Method for Crack Detection

3.1 Structure of the Proposed Method

In this study, sub-dataset generation and matching methods are proposed to improve the performance of CNN-based crack detection on brick walls. The overall structure of the proposed crack detection method consisting of sub-dataset generation and matching is illustrated in Fig. 2 [27]. The mechanism of the sub-dataset generation and the matching methods is shown in Fig. 3 [27].

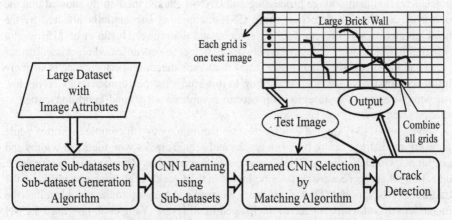

Fig. 2. Structure of the proposed crack detection method (Talukder et al. [27]).

In the proposed method, firstly, a large dataset of images is taken with image attributes. After this, multiple sub-datasets are generated based on the image attributes using the proposed sub-dataset generation method from the large dataset as shown in Fig. 2. Then, CNN learning is done using each sub-dataset. Thereafter, a proper learned CNN is selected by the proposed matching method. The proposed matching method selects a proper learned CNN by matching the attributes of an input image with the attributes of the sub-datasets used for CNN learning, as shown in Fig. 3. Finally, crack detection is performed using the selected proper learned CNN. In general, the perfor-mance of a CNN is low when a large dataset is used for CNN learning containing the images of different types of bricks. Therefore, multiple sub-datasets are generated for CNN learning in which each sub-dataset contains images with a similar type of brick wall. The performance of CNN improved for the crack detection in brick walls by the proposed method because of using a proper sub-dataset for crack detection by matching the test images with the sub-datasets.

From the images in Fig. 1, one can observe that the images of brick walls contain sufficient information on R, G, and B values. Furthermore, the RGB color provides strong indications and increases the image classification accuracy [22, 23]. Therefore, the R, G, and B values of images are considered attributes for the sub-dataset generation and the matching method.

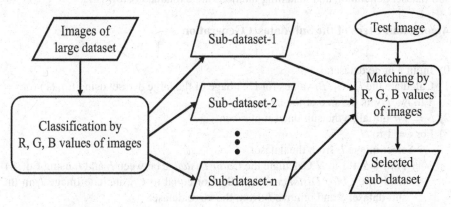

Fig. 3. Mechanism of the sub-dataset generation and matching (Talukder et al. [27]).

3.2 Sub-dataset Generation and Matching

The color distances of the R, G, and B values between the images are calculated to generate small sub-datasets from a large dataset and to match the test input image with the sub-datasets. In the process of sub-dataset generation and matching, the averages of the R, G, and B values of the images must first be calculated. The average R, G, and B values of the ith image are determined using the following Eq. (1):

$$\begin{bmatrix} \overline{R_i} \\ \overline{G_i} \\ \overline{B_i} \end{bmatrix} = \frac{1}{n_x n_y} \sum_{x=1}^{n_x} \sum_{y=1}^{n_y} \begin{bmatrix} r_{xy} \\ g_{xy} \\ b_{xy} \end{bmatrix} \tag{1}$$

In Eq. (1), the spatial coordinates are represented by x and y, image size is represented by n_x and n_y, and the R, G, and B values of the corresponding pixel position in the image are represented by r_{xy}, g_{xy}, and b_{xy}, respectively. If the average R, G, and B values of two images are represented using Eqs. (2) and (3) given below:

$$\{\overline{R_1}, \ \overline{G_1}, \ \overline{B_1}\} \tag{2}$$

$$\{\overline{R_2}, \overline{G_2}, \overline{B_2}\} \tag{3}$$

Then, the *Color Distance* between these two images can be calculated using Eq. (4) shown below:

$$Color\ Distance = \sqrt{\left(\overline{R}_1 - \overline{R}_2\right)^2 + \left(\overline{G}_1 - \overline{G}_2\right)^2 + \left(\overline{B}_1 - \overline{B}_2\right)^2}. \tag{4}$$

The *Color Distance* between zero and C_D was used by the sub-dataset generation method to generate multiple sub-datasets from the large dataset, where C_D was used as the threshold value for the *Color Distance* parameter. The algorithmic steps of the sub-dataset generation and matching methods are explained below [27].

Algorithmic Steps of the Sub-dataset Generation

0) Input a large dataset.
1) Calculate the $\overline{R}_i, \overline{G}_i, \overline{B}_i$ values for i th image of the large dataset using Eq. (1) for $i = 1$ to N. // N is the size of the large dataset.
2) $S = 1$ // Initialize the sub-dataset number.
3) For $i = 1$ to N
   Select image I_i from the dataset.
   3-1) For $j = i + 1$ to N Calculate the *Color Distance* between I_i and I_j using Eq. (4).
        If the *Color Distance* is less than or equal to C_D, include Image I_j in the sub-dataset S and remove I_j from the large dataset.
   3-2) $S = S + 1$.
4) Output the generated sub-datasets.

Algorithmic Steps of the Matching

0) Input sub-datasets and a test image.
1) For $S = 1$ to the number of the sub-datasets
   Calculate the $\overline{\overline{R}_S}, \overline{\overline{G}_S}, \overline{\overline{B}_S}$ values, which are the average values of the $\overline{R}, \overline{G}, \overline{B}$ values of the images of sub-dataset S (example in Fig. 4).
2) Calculate the $\overline{R}, \overline{G}, \overline{B}$ values of the test image using Eq. (1).
3) For $S = 1$ to the number of the sub-datasets
   Calculate the *Color Distance* between the test image and sub-dataset S using Eq. (4).
4) The sub-dataset with the minimum *Color Distance* is selected for the crack detection.

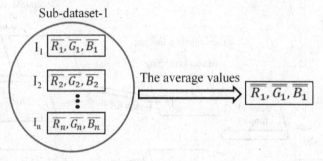

Sub-dataset-1

Fig. 4. Illustrative example of step-1 of matching algorithm for selection method.

The detailed process for selecting a proper sub-dataset by matching the test image with sub-datasets using the matching algorithm is illustrated in Fig. 5.

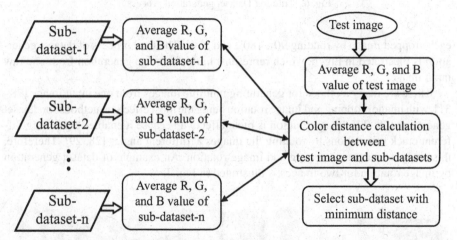

Fig. 5. Selection of proper sub-dataset by matching algorithm.

4   Numerical Experiments

4.1   Dataset Preparation for CNN Learning

The CNN was trained using a total of 1600 images (800 crack images as crack dataset and 800 no-crack images as no crack dataset) of brick walls with a size of 227×227 pixels generated from 100 raw images collected from the Internet. As there are different types of brick walls, images of different bricks were collected from the Internet to cover all bases.

For the training aspect of the dataset generation, firstly, two images are generated from one raw image by manual cropping, after which three images are generated from

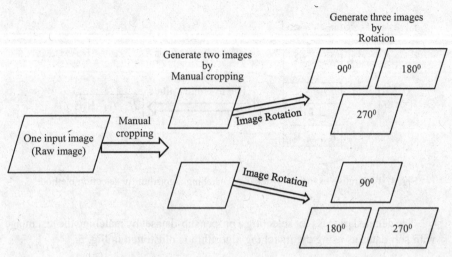

Fig. 6. Training Dataset generation process.

each cropped image by rotating 90°, 180°, and 270°. The total process of dataset generation is illustrated in Fig. 6, which represents multiple image generation from one raw input image.

There are many methods for generating multiple images from one input image [28–31], with image cropping and image rotation being the most effective methods for dataset generation [28, 29]. Image rotation is most effective because it enables examining different crack orientations by rotating the images at different angles [28, 29]. Therefore, this study uses image cropping and image rotation. An example of dataset generation by image rotation for two images is illustrated in Fig. 7.

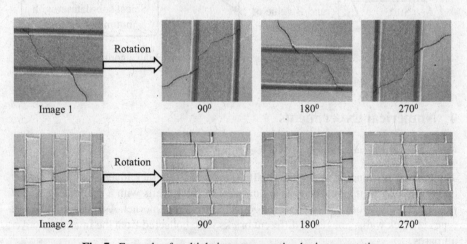

Fig. 7. Example of multiple image generation by image rotation.

4.2 Experimental Design

Crack detection was performed using the proposed method in MATLAB. For the numerical experiments, the CNN architecture was designed with eight layers [5, 13]. Table 1 lists the detailed dimensions of each layer and their operations.

In Table 1, L represents the layers, and C and P represent the convolution and pooling, respectively. For the experiments, the number of epochs was set to 10, and the batch size was 32.

Two experiments—Exp. 1 and Exp. 2—were performed to evaluate the performance of the proposed crack detection method. Exp. 1 evaluated the performance of the CNN for the crack detection in brick walls using a large dataset consisting of the different types of images to train the CNN. Exp. 2 the effectiveness of the sub-dataset generation using the RGB values of images as image attributes and the matching method for the crack detection.

Images with similar RGB values were obtained for each sub-dataset. Subsequently, the generated sub-datasets were used to perform the CNN learning. The properly learned CNN was selected using the matching algorithm and then used to perform the crack detection.

Table 1. Dimensions of the layers and operations of the CNN.

Layers	Height	Width	Depth	Operator	Height	Width	Depth	No	Stride
Input	227	227	3	C1	15	15	3	24	2
L1	107	107	24	P1	7	7	–	–	2
L2	51	51	24	C2	11	11	24	48	2
L3	21	21	48	P2	5	5	–	–	2
L4	9	9	48	C3	9	9	48	96	2
L5	1	1	96	ReLU	–	–	–	–	–
L6	1	1	96	C4	1	1	96	2	1
L7	1	1	2	Softmax	–	–	–	–	–
L8	1	1	2	–	–	–	–	–	–

The results of Exp. 1 and Exp. 2 were obtained by k-fold cross-validation. There are several types of methods for performing cross-validation in machine learning, amongst which the k-fold cross-validation method is the most popular and widely used [32–35]. The k-fold cross-validation procedure has a single parameter called k, referring to the number of groups a given dataset is to be split into [32–35]. In k-fold cross-validation, the original dataset is equally partitioned into k subsets or folds. Out of the k-folds or subsets, for each iteration, one subset is selected as test data (validation), and the remaining ($k-1$) groups are selected as training data. The process is repeated for k times until each group is treated as validation and remaining as training data. The final accuracy of the model is computed by taking the average accuracy of the total validation data.

In the study's experiments, the value of k was set to five, implying that the dataset (1600 images: 800 cracks and 800 no-crack) was divided into five subsets, and testing was done by 5-folds. In each fold, one subset (320 images: 160 cracks and 160 no-crack) was used for testing, with the other four subsets used for training. The summary of the experimental design is shown in Table 2.

Table 2. Summary of the experimental design of Exp. 1 and Exp. 2 for cross-validation.

	Exp. 1	Exp. 2
Purpose	To evaluate the performance of crack detection using a large dataset for training	To evaluate the performance of crack detection using sub-dataset generation and selection methods
Training Dataset	Total 1280 images 1. Crack (640 images) 2. No-crack (640 images)	
Class Label	Crack No-crack	
Sub-dataset generation	No sub-dataset generation	Generate sub-datasets using sub-dataset generation method
Training Dataset Selection	No selection	Select proper sub-dataset using selection method
Test Dataset	320 images (160 cracks, 160 no-crack) at a time for each fold	

For the sub-dataset generation and a proper sub-dataset selection in Exp. 2, the results were checked for different values of C_D ($C_D = 20$ and 40) to select the appropriate threshold value. At $C_D = 20$, the best results were observed. Therefore, the threshold value was set at $C_D = 20$ for Exp. 2; however, this can change with a change in the images of the training dataset.

4.3 Results

In the experiments, four types of data, i.e., true positives, true negatives, false negatives, and false positives, were obtained [36], where positive means there is a crack and negative means there is no crack.

True Positives (TP). TP is the correctly predicted positive value, implying that the actual class is positive, and the predicted class is also positive.

True Negatives (TN). TN is the correctly predicted negative value, implying that the actual class is negative, and the predicted class is also negative.

False Negatives (FN). FN is the incorrectly predicted negative value, implying that the actual class is positive, but the predicted class is negative.

False Positives (FP). FP is the incorrectly predicted positive value, implying that the actual class is negative, but the predicted class is positive.

Four performance metrics—precision, recall, F-measure, and accuracy [13, 36]—were calculated to compare the performances of crack detection in each experiment using the above data.

Precision. The ratio of TPs to the total number of predicted positive observations. High precision is related to a low false-positive rate.

$$Precision = TP/(TP + FP) \tag{5}$$

Recall. The ratio of TPs to all observations in actual positive classes. High recall means the number of false negatives is high.

$$Recall = TP/(TP + FN) \tag{6}$$

F-measure. The weighted average of precision and recall. A high F-measure means that false positives and false negatives are both low.

$$F-measure = (2*Precision*Recall)/(Precision + Recall) \tag{7}$$

Accuracy. The ratio of correctly predicted observation to total observations.

$$Accuracy = (TP + TN)/(TP + FP + FN + TN) \tag{8}$$

When the crack detection method is accurate, the values of precision, recall, F-measure, and accuracy are all close to 1.0, while they are almost 0.0 when the model is improper [36].

The quantitative results of the experiments are presented in Tables 3 and 4, which compare the performances of Exp. 1 and Exp. 2.

Considering Table 3, many false results (FPs and FNs) were obtained in Exp. 1. In contrast, the number of FPs and FNs were reduced in Exp. 2.

From the values of the measurement's metrics obtained by Exp. 1 in Table 4, the range between the minimum and maximum values of the metrics was large: the minimum value of precession was 0.834, whereas the maximum value was 0.917; the minimum value of recall was 0.825, whereas the maximum value was 1.0; the minimum value of the F-measure was 0.842, whereas the maximum value was 0.928; the minimum value of accuracy was 0.841, whereas the maximum value was 0.922. Therefore, Exp. 1 was unstable for different subsets at the time of testing by cross-validation.

In contrast, as shown in Table 4, the range between the minimum and maximum values of the measurement metrics was small for all the subsets in Exp. 2, i.e., the minimum and maximum values of accuracy were 0.978 and 0.991, respectively. Similarly, a small range was observed between the minimum and maximum values for the other measurement metrics for all the subsets. Therefore, the Exp. 2 was stable for all the subsets.

Upon comparing the performance metrics of both experiments, the values of the performance metrics improved in Exp. 2 as compared to Exp. 1, as shown in Table 4. From these results, it is evident that the proposed method can improve the performance of crack detection in brick walls.

Table 3. Summary of the results of Exp. 1 and Exp. 2.

		Exp. 1		Exp. 2	
		Positive (True)	Negative (True)	Positive (True)	Negative (True)
Fold 1	Positive (Estimate)	TP 136	FP 27	TP 157	FP 0
	Negative (Estimate)	FN 24	TN 133	FN 3	TN 160
Fold 2	Positive (Estimate)	TP 160	FP 25	TP 160	FP 4
	Negative (Estimate)	FN 0	TN 135	FN 0	TN 156
Fold 3	Positive (Estimate)	TP 132	FP 12	TP 157	FP 0
	Negative (Estimate)	FN 28	TN 148	FN 3	TN 160
Fold 4	Positive (Estimate)	TP 136	FP 16	TP 158	FP 3
	Negative (Estimate)	FN 24	TN 144	FN 2	TN 157
Fold 5	Positive (Estimate)	TP 152	FP 17	TP 157	FP 4
	Negative (Estimate)	FN 8	TN 143	FN 3	TN 156

Table 4. Comparison of the performance metrics of the experiments.

	Exp. 1				Exp. 2			
	Precession	Recall	F-measure	Accuracy	Precession	Recall	F-measure	Accuracy
Fold 1	0.834	0.850	0.842	0.841	1.0	0.981	0.990	0.991
Fold 2	0.865	1.0	0.928	0.922	0.976	1.0	0.988	0.988
Fold 3	0.917	0.825	0.869	0.875	1.0	0.981	0.990	0.991
Fold 4	0.895	0.850	0.872	0.875	0.981	0.988	0.984	0.984
Fold 5	0.899	0.950	0.924	0.922	0.975	0.981	0.978	0.978
Min:	0.834	0.825	0.842	0.841	0.975	0.981	0.978	0.978
Max:	0.917	1.0	0.928	0.922	1.0	1.0	0.990	0.991
Average:	0.882	0.895	0.887	0.887	0.986	0.986	0.986	0.986

5 Discussion

From the results of Exp. 1 in Table 3, many false results (FPs and FNs) were obtained because some brick structures appeared to have cracks. In some cases, the joining points of bricks also appeared as cracks. To clarify this result, one example of the test image, where Exp. 1 provides a false result, is shown in Fig. 8; however, Exp. 2 provided the correct crack detection result for the image of Fig. 8 by selecting a proper sub-dataset.

Example of Test image

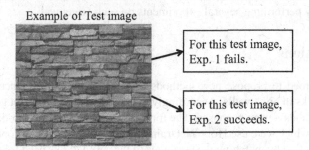

For this test image, Exp. 1 fails.

For this test image, Exp. 2 succeeds.

Fig. 8. Example of test image, where Exp. 1 failed, but Exp. 2 succeeded.

For the image of the brick wall in Fig. 8, Exp. 1 failed because the large dataset contained many different types of color information, i.e., the color of bricks and the joining points of bricks can be different in each image, as shown in Fig. 1, thus making crack detection difficult. Therefore, Exp. 1 failed for the types of images shown in Fig. 8.

Conversely, Exp. 2 was successful for the image of the brick wall in Fig. 8 because the sub-datasets (i.e., a total of eight sub-datasets) were generated, and CNN training was performed using the generated sub-datasets. The learned CNN was selected by matching the test image with the sub-datasets. The selected sub-dataset only contained images similar to the test image; therefore, similarities were found between the test image and the images of this sub-dataset, and thus, Exp. 2 succeeds. The sub-dataset selected for the crack detection of the test image of Fig. 8 is shown in Fig. 9.

Fig. 9. Sub-dataset used for the test image of Fig. 8 by Exp. 2.

By comparing the test image of Fig. 8 and the images of the sub-dataset of Fig. 9, this sub-dataset contains images that are almost similar to the test image shown in Fig. 9. For this reason, Exp. 2 succeeds for this test input image.

The advantage of the proposed method is that the values of the performance metrics are improved for the test images of brick walls. However, a limitation of the method is that the threshold value (C_D) used for the *Color Distance* parameter changes with a change in the images of the training dataset.

The value of C_D should be determined based on the images of the dataset used for the crack detection. If the value of C_D is small, then the number of sub-datasets increases, and the number of images in each sub-datasets decreases. For this reason, a comparatively small value of C_D is better when the dataset is comparatively large. The best value of C_D depends on the image attributes, size of datasets, and the targets of the crack detection. In the practical field of crack detection, the best value of C_D will be determined by performing several experiments on each target wall.

6 Conclusions

This study aimed to propose new methods to improve the performance of CNN for detecting cracks in brick walls. Accordingly, two methods—sub-dataset generation and matching methods—were proposed. The former generated multiple sub-datasets from a large dataset, which were used for CNN training. The latter selected a proper sub-dataset for crack detection by matching the attributes of the sub-datasets used for CNN training with those of the test image. The crack detection performance of CNN for brick walls was evaluated using cross-validation.

As seen in Table 4, the values of the metrics obtained by the cross-validation were increased using Exp. 2, which used proposed sub-dataset generation and matching methods, compared to Exp. 1. In addition, the range between the minimum and maximum values of the metrics was smaller for Exp. 2 compared to Exp. 1. Considering the results of cross-validation, the proposed method improves the performance of crack detection in different types of brick walls due to the generation of multiple sub-datasets for CNN training and for selecting a proper sub-dataset for crack detection.

Proposed method was successful to detect cracks in different types of brick walls. Therefore, this study will help to ensure the safety and durability of structures in real life, as well as ensure the safety of human life.

In this study, the images of the training dataset were prepared by collecting 100 images from the internet. Future research will investigate crack detection by preparing training datasets from real target walls.

In this study, crack detection was performed based on a binary decision, such as crack or no-crack detection; however, judgments of other aspects of the cracks, such as their length, width, or depth, were not considered. Considering these aspects are important to properly understand crack damage (such as identifying heavy damage or light damage) and is one of the major issues for future research.

References

1. Liu, Z., Cao, Y., Wang, Y., Wang, W.: Computer vision-based concrete crack detection using U-net fully convolutional networks. Autom. Constr. **104**, 129–139 (2019)
2. Road Bureau Japan: Road maintenance in Japan: Problems and solutions. Ministry of land, infrastructure, transport and tourism, Roads in Japan (2015)
3. American Society of Civil Engineers (ASCE): Infrastructure Report Card (2017)

4. Dais, D., Bal, I.E., Smyrou, E., Sarhosis, V.: Automatic crack classification and segmentation on masonry surfaces using convolutional neural networks and transfer learning. Autom. Constr. **125**, 1–18 (2021)
5. Cha, Y.J., Choi, W.: Deep learning-based crack damage detection using convolutional neural networks. Comput.-Aided Civil Infrastruct. Eng. **32**, 361–378 (2017)
6. Ozgenel, C.F., Sorguc, A.G.: Performance comparison of pretrained convolutional neural networks on crack detection in buildings. In: International Symposium on Automation and Robotics in Construction, ISARC (2018)
7. Choi, D., Jeon, Y., Lee, S.J., Yun, J.P., Kim, S.W.: Algorithm for detecting seam cracks in steel plates using a Gabor filter combination method. Appl. Opt. **53**(22), 4865–4872 (2014)
8. Neogi, N., Mohanta, D.K., Dutta, P.K.: Review of vision-based steel surface inspection systems. EURASIP J. Image Video Process. **2014**(1), 1–19 (2014). https://doi.org/10.1186/1687-5281-2014-50
9. Qader, I.A., Abudayyeh, O., Kelly, M.: Analysis of edge detection techniques for crack identification in bridges. J. Comput. Civ. Eng. **17**(4), 255–263 (2003)
10. Wu, X., Xu, K., Xu, J.: Application of undecimated wavelet transform to surface defect detection of hot rolled steel plates. In: 2008 Congress on Image and Signal Processing (2008)
11. Yeum, C., Dyke, S.: Vision-based automated crack detection for bridge inspection. Comput.-Aided Civil Infrastruct. Eng. **30**, 759–770 (2015)
12. Talukder, M.H., Ota, S., Takanokura, M., Ishii, N.: Crack detection of concrete walls by CNN using sub-datasets. In: The 2020 Spring National Conference of Operations Research Society of Japan, pp. 82–83, Japan (2020)
13. Talukder, M.H., Ota, S., Takanokura, M., Ishii, N.: Crack detection in concrete structures under varied environmental conditions using CNN. J. Soc. Plant Eng. Japan **33**(1), 14–21 (2021)
14. Dung, C.V., Anh, L.D.: Autonomous concrete crack detection using deep fully convolutional neural network. Autom. Constr. **99**, 52–58 (2019)
15. Huyan, J., Li, W., Tighe, S., Zhai, J., Xu, Z., Chen, Y.: Detection of sealed and unsealed cracks with complex backgrounds using deep convolutional neural network. Autom. Constr. **107**, 1–14 (2019)
16. Jacob, K., Mark, D.J., Peter, B., Mike, M., Gordon, M.: A convolutional neural network for pavement surface crack segmentation using residual connections and attention gating. In: 2019 IEEE International Conference on Image Processing, ICIP (2019)
17. Li, S., Zhao, X.: Image-based concrete crack detection using convolutional neural network and exhaustive search technique. Adv. Civil Eng. **2019**, 1–12 (2019)
18. Li, G., Ma, B., He, S., Ren, X., Liu, Q.: Automatic tunnel crack detection based on u-net and a convolutional neural network with alternately updated clique. Sensors **20**, 1–23 (2020)
19. Mahtab, M.K., et al.: Deep-learning-based crack detection with applications for the structural health monitoring of gas turbines. Struct. Health Monit. **19**(5), 1440–1452 (2019)
20. Hoang, N.D., Nguyen, Q.L., Tran, V.D.: Automatic recognition of asphalt pavement cracks using metaheuristic optimized edge detection algorithms and convolution neural network. Autom. Constr. **94**, 203–213 (2018)
21. Andrushia, A.D., Anand, N., Godwin, I.A.: Analysis of edge detection algorithms for concrete crack detection. Int. J. Mech. Eng. Technol. **9**(11), 689–695 (2018)
22. Bianconi, F., Harvey, R., Southam, P., Fernandez, A.: Theoretical and experimental comparison of different approaches for colour texture classification. J. Electron. Imaging **20**(4), 1–20 (2011)
23. Varma, M., Zisserman, A.: A statistical approach to texture classification from single images. Int. J. Comput. Vision **62**(1), 61–81 (2005)
24. Zhang, K., Cheng, H.D., Zhang, B.: A unified approach to pavement crack and sealed crack detection using preclassification based on transfer learning. J. Comput. Civil Eng. **32** (2018)

25. Zhang, L., Yang, F., Zhang, Y.D., Zhu, Y.J.: Road crack detection using a deep convolutional neural network. In: Proceedings of the International Conference on Image Processing (ICIP), pp. 3708–3712, Phoenix, AZ-USA (2016)
26. Yang, X., Li, H., Yu, Y., Luo, X., Huang, T., Yang, X.: Automatic pixel-level crack detection and measurement using a fully convolutional network. Comput.-Aided Civil Infrastruct. Eng. **33**, 1090–1109 (2018)
27. Talukder, M.H., Ota, S., Takanokura, M., Ishii, N.: Sub-dataset generation and matching for crack detection on brick walls using convolutional neural network. In: Proceedings of 2nd International Conference on Deep Learning Theory and Applications, DeLTA 2021, pp. 191–197, Lisbon-Portugal (Online streaming) (2021)
28. Wang, Z., Yang, J., Jiang, H., Fan, X.: CNN training with twenty samples for crack detection via data augmentation. Sensors **20**(17), 1–17 (2020)
29. Shorten, C., Khoshgoftaar, T.M.: A survey on image data augmentation for deep learning. J. Big Data **6**(1), 1–48 (2019)
30. Takahashi, R., Matsubara, T., Uehara, K.: Data augmentation using random image cropping and patching for deep CNNs. IEEE Trans. Circuits Syst. Video Technol. **30**, 2917–2931 (2015)
31. Takahashi, R., Matsubara, T., Uehara, K.: RICAP: random image cropping and patching data augmentation for deep CNNs. Proc. Mach. Learn. Res. **95**, 786–798 (2018)
32. Berrar, D.: Cross-validation. Encycl. Bioinform. Comput. Biol. **1**, 542–545 (2018)
33. Rawat, A.S.: Introduction to Cross-validation in Machine Learning. Retrieved from https://www.analyticssteps.com/blogs/introduction-cross-validation-machine-learning. Accessed 10 June 2022
34. Brownlee, J.: A Gentle Introduction to k-fold Cross-validation. Retrieved from https://machinelearningmastery.com/k-fold-cross-validation/. Accessed 10 June 2022
35. Mujtaba, H.: What is Cross-validation in Machine Learning? Types of Cross-Validation. Retrieved from https://www.mygreatlearning.com/blog/cross-validation/. Accessed 10 June 2022
36. Baratloo, A., Hosseini, M., Negida, A., Ashal, G.: Simple definition and calculation of accuracy, sensitivity and specificity. Emergency **3**(2), 48–49 (2015)

Author Index

A
Alharbi, Yassir 88
André, Elisabeth 1, 67
Arribas-Bel, Daniel 88

C
Coenen, Frans 88

D
Diab, Lama 67
Dumpala, Veeru 24

F
Ferreira, Bruno Georgevich 49

G
Garcia, Sebastian 111
Geinitz, Steffen 1

I
Ishii, Nobuaki 134

K
Kiderle, Thomas 67
Kroner, Valentin 67
Kurupathi, Sheela Raju 24

L
Lima, Bruno Gabriel Cavalcante 49
Lingenfelser, Florian 67
Lukas, Ondrej 111

M
Margraf, Andreas 1
Mertes, Silvan 1, 67

O
Ota, Shuhei 134

S
Schiller, Dominik 67
Stricker, Didier 24

T
Takanokura, Masato 134
Talukder, Mehedi Hasan 134

V
Vieira, Tiago Figueiredo 49

© The Editor(s) (if applicable) and The Author(s), under exclusive license
to Springer Nature Switzerland AG 2023
A. Fred et al. (Eds.): DeLTA 2020/DeLTA 2021, CCIS 1854, p. 151, 2023.
https://doi.org/10.1007/978-3-031-37320-6

Author Index

Printed in the United States
by Baker & Taylor Publisher Services.

Printed in the United States
by Baker & Taylor Publisher Services